HELMAND, AFGHANISTAN

3COMMANDO BRIGADE

EWEN SOUTHBY-TAILYOUR

EBURY
PRESS

3 5 7 9 10 8 6 4 2

First published in 2008 by Ebury Press, an imprint of Ebury Publishing
A Random House Group company
This edition published 2009

The Random House Group Limited Reg. No. 954009

Addresses for companies within the Random House Group can be found at
www.randomhouse.co.uk

A CIP catalogue record for this book is available from the British Library

The Random House Group Limited supports The Forest Stewardship
Council (FSC), the leading international forest certification organisation.
All our titles that are printed on Greenpeace approved FSC certified paper
carry the FSC logo. Our paper procurement policy can be found at
www.rbooks.co.uk/environment

Typeset by seagulls.net

Printed in the UK by CPI Cox & Wyman, Reading, RG1 8EX

ISBN 9780091926960

To buy books by your favourite authors and register for offers visit
www.rbooks.co.uk

*'The Royal Marines' deployment was among
the finest pieces of soldiering I have come across.'*

General Sir Richard Dannatt, KCB, CBE, MC
Chief of the General Staff

Dedicated to the memory of those members of

3 Commando Brigade, Royal Marines

who were killed during *Operation Herrick 5*
in Helmand Province, Afghanistan
September 2006 to April 2007

Warrant Officer Class 2 Michael Smith
29 Commando Regiment, Royal Artillery
Killed during a grenade attack on the UK base in Sangin
Thursday 8 March 2007 – aged thirty-nine

Marine Benjamin Reddy
42 Commando Royal Marines
Killed while attacking enemy positions in Kajaki
Tuesday 6 March 2007 – aged twenty-two

Lance Bombardier Ross Clark
29 Commando Regiment, Royal Artillery
Killed during a rocket attack on the UK base in Sangin
Saturday 3 March 2007 – aged twenty-five

Lance Bombardier Liam McLaughlin
29 Commando Regiment, Royal Artillery
Killed during a rocket attack on the UK base in Sangin
Saturday 3 March 2007 – aged twenty-one

Marine Scott Summers
42 Commando, Royal Marines
Died following a road traffic accident
Wednesday 21 February 2007 – aged twenty-three

Marine Jonathan Holland
45 Commando, Royal Marines
Killed by an anti-personnel mine in Sangin
Wednesday 21 February 2007 – aged twenty-three

Lance Corporal Mathew Ford
45 Commando, Royal Marines
Killed while attacking Jugroom Fort
Monday 15 January 2007 – aged thirty

Marine Thomas Curry
42 Commando, Royal Marines
Killed while attacking enemy positions at Kajaki
Saturday 13 January 2007 – aged twenty-one

Lance Bombardier James Dwyer
29 Commando Regiment, Royal Artillery
Killed when the vehicle he was driving struck an
anti-tank mine in southern Helmand
Wednesday 27 December 2006 – aged twenty-two

Marine Richard Watson
42 Commando, Royal Marines
Killed in an ambush at Now Zad
Tuesday 12 December 2006 – aged twenty-three

Marine Jonathan Wigley
45 Commando, Royal Marines
Killed while attacking enemy positions at Garmsir
Tuesday 5 December 2006 – aged twenty-one

Marine Gary Wright
45 Commando, Royal Marines
Killed by a suicide-borne improvised explosive device in Lashkar Gah
Friday 19 October 2006 – aged twenty-two

CONTENTS

FOREWORD

HRH PRINCE PHILIP

One of the most challenging tasks for a military historian is to give a clear and comprehensible account of a battle. As the story of 3 Commando Brigade in Afghanistan is a series of more or less violent encounters with heavily armed insurgents, the problem of producing a coherent story is all the greater. Just to make matters more complicated, modern warfare demands an extensive new vocabulary that is virtually incomprehensible to the layman.

In this masterly account of the very specialised warfare in Afghanistan, and the very special achievements of 3 Commando Brigade, the author demonstrates his rare ability to explain what went on in terms that any reasonably intelligent layman can comprehend. To say that the story is 'action-packed' is an understatement. That the Royal Marine Commandos were able to sustain such an intense level of action in seriously uncomfortable circumstances is a tribute to their long and arduous training. The book is an eloquent accolade to everyone who took part in this deployment.

Philip

NOTE

3 COMMANDO BRIGADE BATTLE GROUP

The 3rd Commando Brigade, Royal Marines, deployed to southern Afghanistan as the core of the United Kingdom Task Force. The main area of operations was Helmand Province. The UKTF was formed around the following key components:

Helmand Battle Group

42 Commando numbering up to 1,500 men and women.

Command Company with the ISTAR group (Recce Troop and Manoeuvre Support Group), Mortar Troop with 23 mortar barrels instead of 9 and the Assault Engineer Troop.

Juliet, Kilo, Lima and Mike combat companies.

A logistic company.

Plus the following under command:

One company from the army: either the Royal Regiment of Fusiliers or The Rifles.

Whiskey Company, 45 Commando: Force Protection Company for 28 Engineer Regiment.

A battery from 29 Commando Regiment.

C Squadron Light Dragoons: three troops with a mixture of Scimitars, Spartan and Samaritan as an ambulance.

59 Independent Commando Squadron, Royal Engineers.

42 UAV Battery, Royal Artillery.

Estonian Armoured Infantry Company.

Danish Formation Recce Squadron.

The following were attached to the commando battle group at times:

Engineer Group: 28 Engineer Regiment, Royal Engineers provided combat engineer support, a military construction force and reconstruction elements.

Operational Mentoring and Liaison Teams (OMLT): framed on 45 Commando: key in the training, support, development and operations of the Afghan National Army 3 Brigade of the Afghan 205 Corps.

29 Commando Regiment, Royal Artillery: provided six 105 Light Guns and Fire Support Teams across the area of operations.

UKLF (UK Land Force) Command Support Group (CSG): effectively, a second battle group.

Main duties:

Command and communications across the brigade.

Supplying an IX staff cell with commanding officer UKLF CSG as chief information officer. This cell had 2 key roles:

- The coordination of the 'find' function; exploiting the resources of the Brigade Reconnaissance Force, Light Dragoons Squadron, Y Squadron (electronic warfare and counter measures), tactical unmanned aerial vehicle (UAV) and national and coalition strategic and intelligence assets.
- The exploitation of all information to support operations through intelligence assessment and information operations.

India Company: a composite company from 45 Commando Reconnaissance Troop, Air Defence Troop and other UKLF CSG ranks that formed the Lashkar Gah Operations Company.

Zulu Company, 45 Commando.

Commando Logistic Regiment
The CLR provided 2nd and 3rd line support to the UKTF; split between Kandahar and Helmand but based at Camp Bastion. Included its own Convoy Force Protection Troop.

Other Force Components
Explosive Ordnance Disposal Group.
Civil-military cooperation (CIMIC) and PsyOps teams.
Development and influence teams (DITs): a 3 Commando Brigade creation based on Engineer Reconnaissance, CIMIC, PsyOps and Force Protection, delivering the research and development 'find' function.
Joint Provincial and Joint District Coordination Centre Teams (JPCC – JDCC Teams): small bands of advisors seeking to tie together the ANA, ANP and Afghan Intelligence Services.
Provincial Reconstruction Teams (PRTs).

LIST OF MAPS

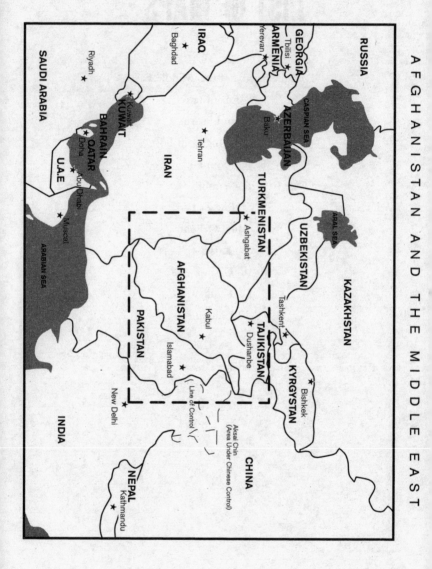

AFGHANISTAN AND THE MIDDLE EAST

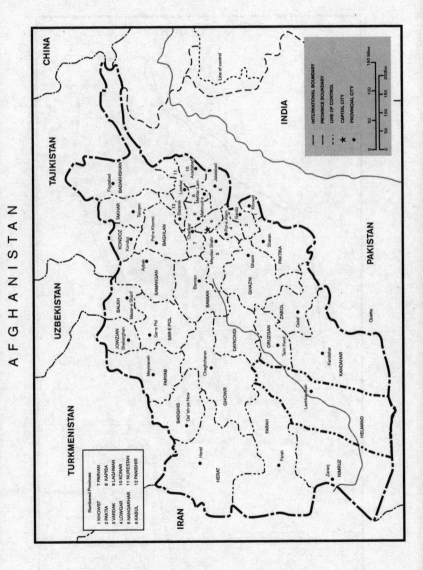

AFGHANISTAN

HELMAND PROVINCE

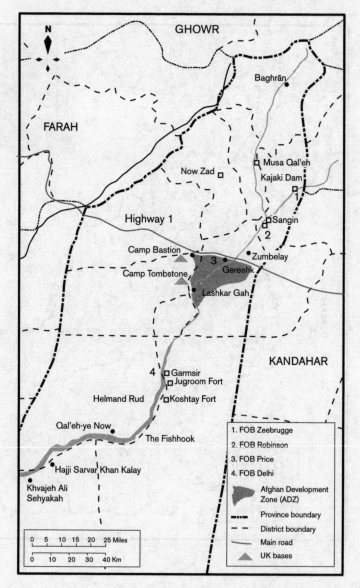

GHOWR

FARAH

Baghrān

Now Zad

Musa Qal'eh

Kajaki Dam

1

Highway 1

Sangin

2

Camp Bastion

3

Zumbelay

Camp Tombstone

Gereshk

Lashkar Gah

KANDAHAR

4   Garmsir

Jugroom Fort

Helmand Rud

Koshtay Fort

Qal'eh-ye Now

The Fishhook

Hajji Sarvar Khan Kalay

Khvajeh Ali
Sehyakah

1. FOB Zeebrugge
2. FOB Robinson
3. FOB Price
4. FOB Delhi

Afghan Development
Zone (ADZ)

Province boundary

District boundary

Main road

UK bases

0 5 10 15 20 25 Miles

0 10 20 30 40 Km

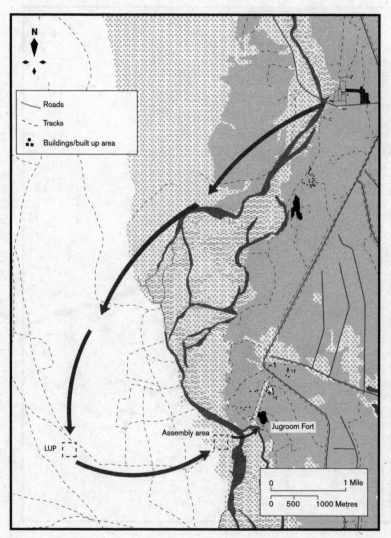

N

Roads
Tracks
Buildings/built up area

Assembly area

LUP

Jugroom Fort

0   1 Mile
0   500   1000 Metres

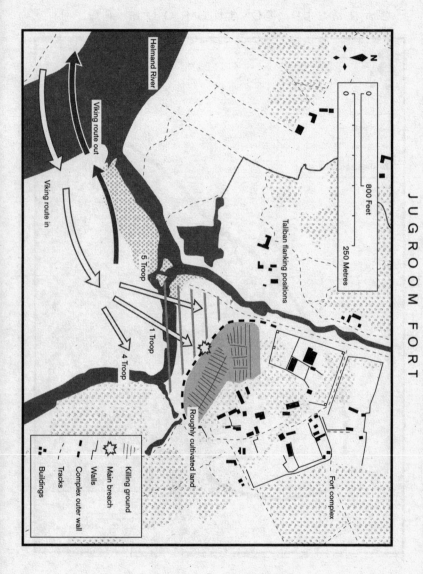

JUGROOM FORT

Helmand River

Viking route out

Viking route in

Taliban flanking positions

5 Troop

1 Troop

4 Troop

Roughly cultivated land

Fort complex

N

800 Feet

250 Metres

Killing ground
Main breach
Walls
Complex outer wall
Tracks
Buildings

GERESHK

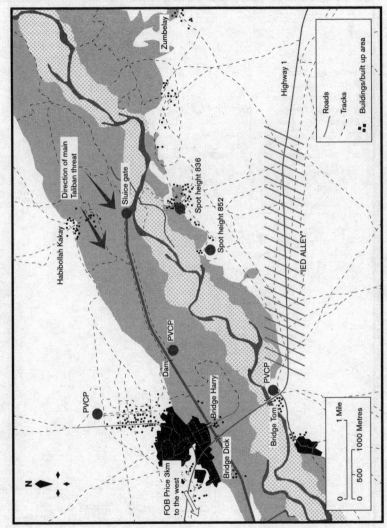

Habibollah Kakay

Direction of main
Taliban threat

Sluice gate

Zumbelay

Spot height 836

Spot height 852

"IED ALLEY"

Highway 1

PVCP

Dam

PVCP

Bridge Harry

Bridge Tom

Bridge Dick

FOB Price 3km
to the west

PVCP

N

Roads
Tracks
Buildings/built up area

1 Mile

0   500   1000 Metres

NOW ZAD

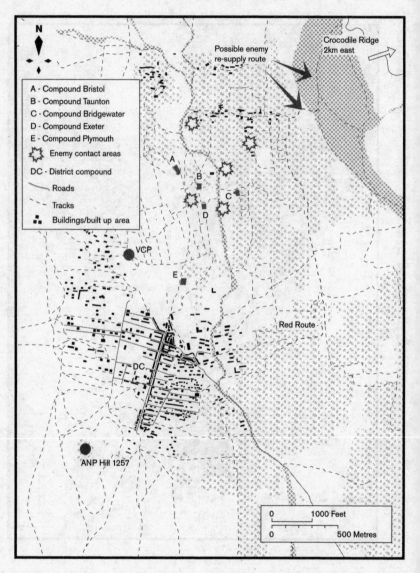

N

A - Compound Bristol
B - Compound Taunton
C - Compound Bridgewater
D - Compound Exeter
E - Compound Plymouth
Enemy contact areas
DC - District compound
Roads
Tracks
Buildings/built up area

Possible enemy re-supply route

Crocodile Ridge 2km east

A
B
C
D

VCP

E

L

Red Route

DC

ANP Hill 1257

0   1000 Feet
0   500 Metres

MUSA QAL'EH

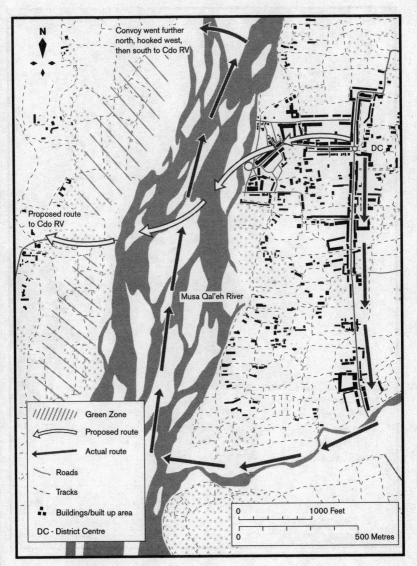

N

Convoy went further
north, hooked west,
then south to Cdo RV

Proposed route
to Cdo RV

DC

Musa Qal'eh River

///////// Green Zone

Proposed route

Actual route

Roads

Tracks

Buildings/built up area

DC - District Centre

0			1000 Feet

0		500 Metres

KAJAKI

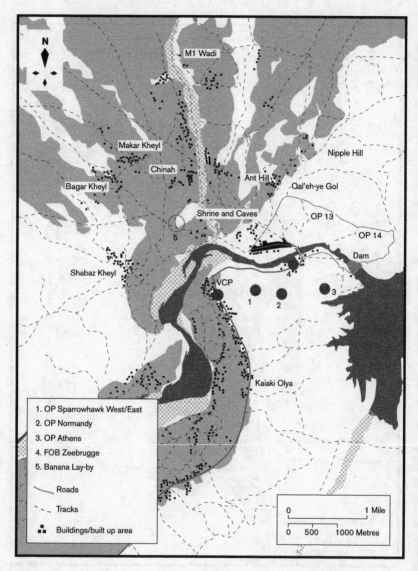

1. OP Sparrowhawk West/East
2. OP Normandy
3. OP Athens
4. FOB Zeebrugge
5. Banana Lay-by

—— Roads

--- Tracks

∴ Buildings/built up area

| 0 | | 1 Mile |
| 0 | 500 | 1000 Metres |

GARMSIR

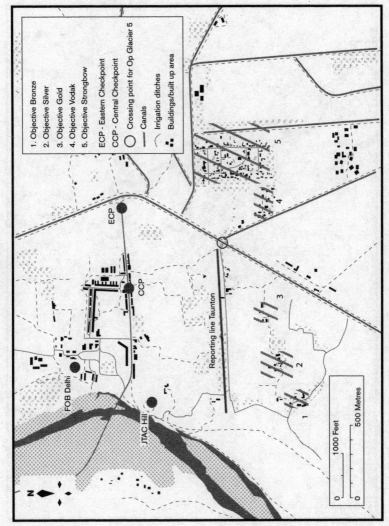

Legend:
1. Objective Bronze
2. Objective Silver
3. Objective Gold
4. Objective Vodak
5. Objective Strongbow

ECP - Eastern Checkpoint
CCP - Central Checkpoint
⭕ Crossing point for Op Glacier 5
— Canals
┄ Irrigation ditches
▪ Buildings/built up area

FOB Delhi
JTAC Hill
ECP
CCP
Reporting line Taunton

1000 Feet
500 Metres

N

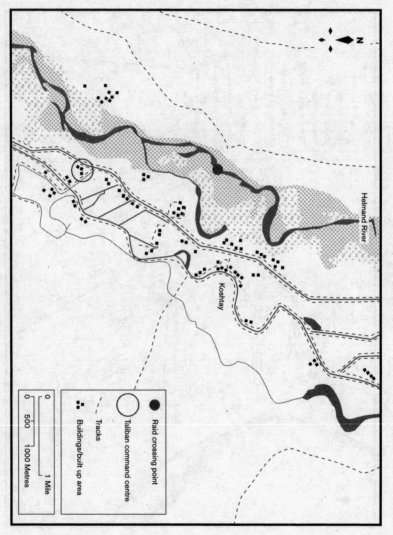

K O S H T A Y

Helmand River

Koshtay

Raid crossing point
Taliban command centre
- - - Tracks
■■ Buildings/built up area

0
500
1000 Metres

0
1 Mile

S A N G I N

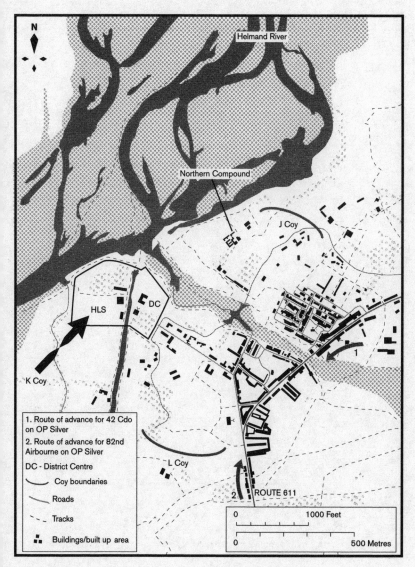

N

Helmand River

Northern Compound

J Coy

DC

HLS

K Coy

1

L Coy

1. Route of advance for 42 Cdo on OP Silver
2. Route of advance for 82nd Airbourne on OP Silver

DC - District Centre

⌒ Coy boundaries

Roads

Tracks

Buildings/built up area

2 ROUTE 611

0   1000 Feet

0   500 Metres

PROLOGUE

JUGROOM FORT – *OPERATION GLACIER TWO*

'It was starting to get light. I remember the water coming halfway up the little window at the back of the Viking. Because of the smoke and dust the Vikings behind lost sight of us so we were static and on our own. Someone came over the net saying he could see at least eight enemy so I said, "Right, lads, let's get the bayonets on."'

Helmand Province is an area roughly half the size of England, with a population about the same as the county of Devon. The temperature during the summer months might exceed 45° Centigrade; in winter it can drop as low as -15°. Divided into two by the Helmand River, which irrigates the land and supplies electricity, the area's chief export is opium, with Helmand believed by many to be the world's largest producer of the raw material of the drug.

In the south of the province lies the area known as Garmsir; close by the eastern bank of the Helmand River. Strung out at the south-western end of village-style compounds, stands Jugroom Fort, a rambling complex with heavily armed watchtowers, known as an important Taliban stronghold. It was a vital command and control centre and, possibly, a sizeable store for enemy munitions. This near-impregnable fort was surrounded by a high, massively thick mud-and-rubble wall, within which were places of worship, walled accommodation compounds, kitchens, stores, magazines and small, cultivated fields divided by low walls.

In the weeks preceding *Operation Glacier Two*, reconnaissance parties had established that the fort was occupied and that the Taliban intended to keep it that way. From their mobile bases in the desert surrounding the fort out to the west, C Squadron under Major Ben Warrack and his Scimitars, together with Captain Jason Milne and the Brigade Patrol Troop running their weapons-mounted Land Rovers, or WMIKS, had come close enough in to have drawn fire from both small arms and rocket-propelled grenades. Being shot at, under the terms of the Rules of Engagement then in force, meant they were allowed to return fire, an action that allowed the commandos to assess the enemy's strength and determination. 'Once the Taliban opened fire we could "mallet" them as hard as we liked.' Marine Adam Edwards had been sent out with 1 Troop to do exactly that about a month or so before the assault took place: 'We saw the pre-dug, stand-to posts and watched two guys preparing a heavy machine gun position. Our orders were that if we saw more than eight Taliban with weapons we could engage – just for a bit of a stand-off. So, when we saw eight guys we started with Javelin to initiate the contact and then all the troop weapons plus the under-slung grenade launchers and everything else was let off. We told the boss that the target was huge – bigger than Ben Hur.'

Apache attack helicopters and Nimrod MR2 aircraft also flew over the fort, adding to the on-the-spot intelligence provided by C Squadron and the Brigade Patrol Troop, confirming that Jugroom was indeed a target for a commando raid.

Once it was decided to launch an assault, it was then a question of how, as the terrain was a defender's dream. The hundred-metre-wide Helmand River ran in towards the fort from the north-west before turning south about 300 metres from the fort's south-western walls. When not in spate the river was fordable on foot, although only awkwardly and slowly, while unwaterproofed, civilian vehicles had been observed wading. Earlier reconnaissance proved that for some distance either side of Jugroom the banks of the river were steep; if the river were to be crossed it would need to be in the amphibious Vikings, and at exactly the position that was least desirable for an attacking force.

As if that were not enough, the whole area around the fort was surrounded and criss-crossed by miles of irrigation channels – small canals – that made access for the uninvited even more difficult. A sandbar separates two tributaries that flow into the river from along the fort's flanks, a confluence that forms a triangle of roughly culti-vated land with the wall of the fort the third side. One marine's view was that it was 'really rubbishy, shitty, muddy, boggy, flat ground'. As well as the terrain, another uncertainty was the possi-ble response of the people in the surrounding area to an attack; would they throw their lot in with the enemy and fight, or merge into the darkness?

There was a third unknown factor that was far more of a prob-lem. The reconnaissance and fly-bys hadn't revealed that Jugroom Fort housed a maze of deep, extensive tunnels in which a large number of men could shelter from all but a direct hit. The Taliban covered the entrances to the tunnels with scrub, and sprayed the branches with water to stop them showing up on thermal-imaging equipment. These warrens didn't just provide the Taliban with a hiding place; as they led from the fort to the compounds outside, they also allowed the enemy to enter and leave unobserved. Not having this information was to expose the commandos to the possi-bility of a surprise counter-attack launched from where they least expected it – beside them and behind them.

Unaware of this potential trap, Colonel Robert Magowan, the commanding officer of the Information Exploitation Group, devised a straightforward plan for *Operation Glacier Two*: 'The aim of the assault was simple and, in accord with other deliberate attacks, was to "disrupt and harass the Taliban on his own ground, to raid and not occupy, to get in fast and get out fast". The plan was to conduct a five-hour bombardment through the night using artillery, mortars, B-1B bombers, F-18 Hornets and Apache attack helicopters to destroy the fort and its towers and when the conditions were right – I had "eyes in the sky" the whole time – I would follow with a pre-dawn assault using 45 Commando's Zulu Company. Even if we had destroyed much of the fort I did not want the enemy to think we were not prepared to actually enter their territory as well: an important psychological factor. My marines would cross the river in Vikings to seize the fort and destroy any remaining Taliban and their *materiel*. When satisfied that none of either was left the marines would return whence they came, back to the comparative sanctuary of the western desert to regroup and prepare for the next operation.'

Captain Chris Witts, the officer commanding 1st Armoured Support Troop, whose Vikings were to be a major key to any success or failure of the ground attack, had been warned that he might have to cross the river. Keen not to do so without a proper reconnaissance, he had earlier sent Sergeant Jason Wood – who was trained in conducting river reconnaissance – to look for a suitable place to the south, out of range of the Taliban's defending weapons. The river, however, narrowed between steep banks and flowed considerably faster when away from the fort: after two days Jason Wood returned without having found a place to cross. As a result Chris Witts confided to his diary that the Vikings, with Zulu Company aboard, would be expected to ford the river '200 metres from known enemy positions'; a manoeuvre with which he was not happy.

In truth there were not enough British forces on the ground in Helmand Province to occupy a fort such as Jugroom and then hold it from the enemy; so the operational tactic was to prevent the Taliban from settling into any base and then resupplying from it, by demonstrating that they were vulnerable to an attack from the British at any time. It was vital to impress upon the Taliban that they were not safe inside any compound, and especially a key one such as Jugroom. By depleting their forces, destroying their munitions and other resources, the Royal Marines were there to give the enemy a bloody nose, and to let them know they could do it again, any time they wanted.

To prepare for his planned assault, on 13 January, Colonel Magowan established his TAC HQ, gun line and lying-up position (LUP) due west of the fort, out of range of prying eyes and RPGs. He then despatched Zulu Company to join the Brigade Reconnaissance Force and C Squadron in the western desert, along with three 105mm guns from 29 Commando Light Regiment, Royal Artillery.

Following standard procedure, a Rehearsal of Concept (ROC) was held around a huge relief model in the sand, the size of a football pitch. This 'walk-through' preparation was attended by commanders at all levels, during which most of the planned manoeuvre details were thrashed out. Fuller rehearsals followed before the company moved in the dark to an Assembly Area, closer to the river. By 0400 the next morning, all was in place should the order to assault be given.

At the LUP the colonel gave his orders, followed at 0900 on 14 January, by those of Zulu Company commander to his troop and section commanders. The first phase of *Operation Glacier Two* was for bombardment of the fort, scheduled for that night; the second phase was the pre-dawn infantry assault.

Captain Dave Rigg, an army officer on the colonel's staff for this operation, was commando-trained; he was also the regimental operations officer of 28 Engineer Regiment, an invaluable part of the brigade's order of battle. Rigg was to spend much of the following night watching the bombardment. 'Crystal-clear footage from a Nimrod flying at altitude showed the fort and its occupants; so we knew where the defences were and had a good idea of how many people were in there and with what weapons. We watched people – little black spots – jog round the fort and disappear into trenches and fire positions.'

Meanwhile the bombardment – including a 'huge amount of

ordnance' from the Scimitars and Zulu Company's Fire Support Troop – was pulverizing targets identified via the Nimrod video footage until about 0300 on 15 January, by when the colonel was confident that the enemy had been suppressed sufficiently for him to launch his troops.

To make the assault easier, Captain Rigg wanted the bombers to remove some of the defensive barriers as well as the Taliban themselves. 'I said to the B-1Bs, "I want a bomb to allow the troops to enter the fort and this is the grid. A 2,000-pounder, please." The B-1B bombers dropped twenty 2,000-pound bombs with one landing precisely where we wanted it, breaching the wall in the south-west of the fort. It was probably too good, as it left a vast crater about twenty feet across and eight feet deep, in fact a bit of an obstacle for the boys.'

Colonel Magowan studied the video feed from the Nimrod alongside Captain Rigg, and felt that the time was right. 'After dropping thousands of pounds of bombs I knew it was time to exploit. As I could see no more enemy I called in Zulu Company commander at 0400 and told him, "*Go*."'

By the time the company commander had readied his men for the river crossing and they were at last prepared, dawn was about to break. Sergeant 'Kenny' Everett of 4 Troop was sitting in his Viking at 0530 when the troop officers were summoned for confirmatory orders; Second Lieutenant Hammond returned to give three minutes of confirmatory orders whose brevity 'was not his fault as that was all the briefing he himself had had from the company commander'.

Now the men of Zulu Company knew that the order of advance would be 5 Troop, then 1 Troop followed by 4 Troop remaining in reserve on a spit of sand before the second, minor, river. A revised H Hour – an original one had been at 0530 – was confirmed as being 'as soon as possible'. The OC was shouting at the officers, 'Hurry up as we need to move – now!' The time was 0550.

Chris Witts remained unenthusiastic about crossing this river within 200 metres of a defended fort, but his Vikings were new to battle and he was desperate that they should do all that was asked of them. Now his received orders were unequivocal: the attack would take place as planned and, although thrilled, he was also anxious. He needn't have been as he watched the leading Viking commander, Corporal Paul 'Tiny' McShannon, forge across the river without hesitation, climb the far bank and head straight for the breach in the fort's wall.

Corporal Al Weldon, inside the Viking commanded by Corporal McShannon, instructed his men to fix bayonets as they advanced towards the fort. 'We lit a last fag as most of us had started smoking: there was hardly anyone who did not smoke that day. When we de-bussed there was an eerie silence in the smoke or mist and in a split second I saw the break-in point and shouted, "Seen!" There was a small house and rubble and to my half-left a high-rise building which was part of the village with large holes in the wall where there should have been windows. The ground was open with irrigation ditches: the worst type because there was not enough cover and yet it was very rough to run across quickly.'

Precious time had been lost. Everyone watching on the video screen beamed from the high-altitude Nimrod into the IX Group's TAC HQ had expected the company to be engaged as it crossed the river in the near-dawn light, but nothing happened. Then, with 5 Troop's Vikings within fifty metres of the walls, the Taliban, most of whom had moved into the flanking compounds outside the fort, did open fire. The enemy must have waited out the bombardment underground, and then moved through the tunnels to the outside compounds in enough numbers to pour fire on to the approaching commandos from every possible location around them, except for the small point of entry the Vikings had made across the river. They had spread out across an arc of nearly 270 degrees, turning the whole triangle of roughly cultivated crops into an instant killing ground. Every one of the Royal Marines had been under fire before – there

7

were no novices in Zulu Company – but no one had ever come under such an intense concentration of small arms and grenades.

'I remember looking at the fort and watching the rounds and RPGs coming from three sides and thinking "fuckin' 'ell," said Company Sergeant Major Steven Shepherd. 'If you were on the left side of a Viking you were being shot at, but when you crawled round to the right you were still being shot at. There were people shouting for support as it was almost every man for himself and slightly chaotic. The enemy must have been in deep bunkers throughout the night as we obviously hadn't killed them all. Now the forward troops were moving into the fort. With our own machine guns and the enemy firing small arms and RPGs, the noise was horrific and we couldn't hear anything over the radio.'

With the enemy fire so fierce, CSM Shepherd felt as vulnerable lying down as moving around the killing ground, so he walked 'as calmly as he could' across the scrub, 'encouraging his lads' while noticing 5 Troop move forward under Lieutenant Hammond's command. Mr Shepherd's critical and experienced eye watched approvingly as 1 Troop 'de-bussed' beneath the fort's walls to take up their supporting fire positions in the dirt.

The lead troop, experiencing the most withering fire as they approached the breach, were relying on support from 1 Troop, lying in the open among their Vikings, firing back at the Taliban; although they did little to diminish the ferocity of enemy rounds and rockets. It was impossible for the CSM to tell how many enemy there were, but 'as one of the troop sergeants ran towards me I fired over his shoulder at a geezer who had popped his head over the fort wall. The sergeant thought I was shooting at him and shouted at me to stop!'

One lucky escape followed another. 'An RPG bounced off the ground before landing on Marine Matty Corcoran's helmet; knocked him out for a bit but he soon recovered and rejoined the fight.' The RPGs were landing so close around them that after one explosive blast CSM Shepherd bellowed at a marine to 'bloody check your back-arcs next time you fire your anti-tank weapon'. The reply was, 'Not me, Sergeant Major, but a fucking RPG.' It had detonated on the ground between them.

The bullets and RPGs continued to spray around the marines who were now too near to the Taliban to call in air support: this was a close-quarters battle such as Zulu Company had never experienced before during the current campaign. 'Even my company clerk fired 500 rounds that morning,' said CSM Shepherd. The bayonets fixed earlier in the safety of the Vikings flickered and gleamed in the tracer light, a chilling statement of intent.

Corporals Weldon and 'Derbs' Derbshire, 5 Troop section commanders, pepper-potted their sections towards the breach. Their orders were to split once inside, Derbs to the right and Al Weldon sharp left, two sections of just six marines each, for their gunners had been seconded to the Fire Support Troop. Corporal Weldon was expecting to find a building that needed clearing but 'as it was still quite dark we were doing it slightly by touch. Once we had captured the buildings, to show them we were not scared of taking the war into their very heart, we were then to fuck off back.'

Inside the fort, for a few brief moments, there was a dip in the noise level as each side tried to determine where the other was. Corporal Weldon manoeuvred alongside the wall, with Marine Spiwak alongside him, both readying themselves to leave the fort once they had destroyed the building they'd been tasked to breach. 'Marine Mike Cleary tossed in a grenade and as soon as he did that the breath was taken out of all of our lungs by the sudden and sheer weight of new fire that erupted from enemy positions within the fort. Small-arms rounds and RPG trails all over the place...then I was shot in the back but I don't know what hit me as there was so much shit flying around.'

Al Weldon was fortunate, for the round that injured him went through his webbing, shattering the Bowman radio before fragmenting into his back as shrapnel. Parts of it are still embedded in his body. At about the same time his troop officer, Matt Hammond, had an even luckier escape when a Russian-made 7.62 short bullet hit a fighting knife that he had been sent by his parents: the bullet, bending the knife in half, was deflected away from his body.

The intense rate of incoming fire continued for about ten more minutes before the first cry of 'Man down – man down' was shouted over the troop radio net. With casualties now being taken, CSM Shepherd – a hefty and experienced warrior with a deep, caring compassion for his men – knew that his priority, indeed his primary task, along with ammunition resupply, was to get the wounded out of the area as swiftly as possible. To do this he needed a Viking. He had already formed firm views on the use of this new beast: 'It is a battlefield taxi, not an armoured fighting vehicle.' Even as a taxi it lacked something for him: 'I am a big man, I couldn't get in and out with my helmet on.'

However, CSM Shepherd had nothing but praise for the Viking crews and their determination to support their fellow Royal Marines: 'The newly qualified lads were superb; all you ever had to do was give them an eight-figure grid and they would be there – they could get right up to the fight and that was essential for ammo and casualties. As a vehicle to bring up ammunition, to take casualties back and to carry our men across difficult country it is brilliant, but it isn't a Warrior and so it should not be used to close an enemy position.' Yet despite its virtues as an amphibious vehicle, the CSM remained unconvinced of its usefulness in a combat situation: 'I never, never want to get out of one of those things in contact again as they are not APCs, although one did feel safe in them. Except for crossing the river – a fairly big river. Al Weldon agreed with what I was thinking on the way across with the water halfway up the door – I am not going to get out of this if there is a problem. On the way back I wasn't so worried.'

Shepherd now had four wounded, 'Al Weldon and three lads from Derbs's section. So I shouted to the men in the fort to get out with their casualties. We lifted them into the backs of two or three vehicles.'

Before they left, CSM Shepherd ran to the front of the lead Viking, climbed up the bar armour and shouted at the driver, 'Make sure you fucking come back – know what I mean?'

'Yes, Sergeant Major!'

'Good lad.'

Tiny McShannon, the Viking commander, needed no orders to return: a brave and intelligent man, he had already assessed the situation for himself and was to cross the river a number of times that dawn.

Concern for his own wounded is an emotion that the British fighting man possesses and the Taliban don't. 'We almost stop the battle to make sure we can save a life if it is saveable,' said one. But the Taliban don't see it that way, and if one of them falls, then no one will come to his aid. 'The enemy were not as good soldiers as us but they didn't mind being killed and that counts for a lot.'

Having attended to his casualties, CSM Shepherd strode across to the company commander's Viking, opened the door and roared above the cacophony of 'crack and thump', 'What's going on, sir?'

'Well, what do you reckon, Sergeant Major?' The major had to shout to make himself heard above the noise of the bullets and RPGs buzzing and cracking nearby.

Shepherd bellowed back, 'We are sitting in the middle of a killing ground and we need to get out of here – either forward or back – that's what I reckon.' One of the chief aims of the operation had been to enter the fort and destroy anything thought to be crucial or of value to the Taliban; that had been accomplished.

Now all they were doing was providing targets for the remaining enemy to shoot at.

Communication on the ground may have been made difficult by the noise that prevented the marines using their radios effectively, but within the Vikings it was possible, at least, to communicate with the CO at his desert LUP. Signaller Corporal Robbie Grey was doing an outstanding job, keeping the battle group informed, stage-managing the evacuation of the casualties and all the while passing back Sitreps to the higher formation. His was the only link between the battle and the colonel and so was relaying every message about the situation and the company commander's views and decisions – 'a super-cool guy'.

The message received back by the LUP was that at this crucial stage in the battle it was clear that all was not going quite according to plan. 'The CO asked me what I thought of the situation and I said I thought it was untenable,' said Chris Witts. 'I grabbed the company commander and had a quick discussion with him and the decision was made to extract. I passed that down to all the troops on the ground but it took about twenty more minutes of engaging the enemy to get the guys into the vehicles because at the time we didn't have any direct method of communicating with the lads. Comms were very difficult but eventually we did get everyone back, in no particular order as it was a question of "open the vehicle doors and jump in". As many guys as we could in every Viking.'

CSM Shepherd supervised the loading of wounded men into the Vikings, juggling in his mind the numbers that needed to be carried back with those still to keep suppressing fire aimed at the Taliban: 'I knew who was wounded and I knew we did not have enough vehicles to get us back as some of those who had taken the wounded had yet to return. At this stage I had 1 Troop in their vehicles, 5 Troop moving out and 4 Troop still on the spit of sand.'

Colonel Magowan, at the LUP, supported the decision to return. 'Having exploited as far as he could under effective enemy fire, the company commander wanted to withdraw.' Which, perhaps, was hardly surprising as the troops had taken more than the expected number of casualties, including a T1 – classified as life-threatening – a couple of T2s – can't self-help – and a couple of T3s. T4, so far unreported, would have been dead. The colonel 'discussed the situation with the company commander' and told him, 'You are the man on the ground; you make the call but you haven't been there very long – about forty-five minutes.' The company commander confirmed his position and so the colonel said 'OK, fine,' and readied the Apaches and artillery to help with the withdrawal. Studying the video, Magowan was able to watch as the vehicles transported the wounded to safety before the Royal Marines could break contact with the enemy. 'They took an awfully long time to turn round but eventually the company commander did get his men together and as soon as they were across the river and out of contact the firing stopped.'

However, before the last man had left the killing area one marine observed that the enemy's bottle 'was getting stronger and stronger and although no one was actually charging towards us they were all in firing positions aiming at us'. It was clearly time to move, but as they were leaping on board a Viking, 'Spinner' Spence ran up to the company sergeant major shouting, 'I'm missing a bloke.'

CSM Shepherd asked 'where he could be', and Spinner replied, 'One of the lads thinks he went back in an earlier vehicle as he was with some of the casualties.' The CSM was initially happy with that but 'thought I should speak on the net to check'. It was the first time for some minutes that he could hear anything over the radio – just enough – as RPGs were exploding all around them, and the small-arms fire was still buzzing past. 'I got on the net and because of the din I didn't get the Troop Bible out as I would not be able to make any numbers heard over the air. I simply said, "Have you got Lance Corporal Ford?" and they came back from the Assembly Area and said, "Yeah, he's here." As we only had one Ford in the company I thought that they had him.

'We were still taking incoming as we hadn't moved from the killing ground and as I was alongside 5 Troop in their vehicle, I said, "We'll have one final check," so "Spinner" on his own ran off to the edge of the fort wall where they had all been and returned via the break-in hole before running along the outside wall looking for him.'

He sprinted back to the Viking, waving as he did so, making it clear there was no sign of anyone left behind. He then clambered on to the vehicle as it drove up to the wall, with some men sitting on the bonnet. Shepherd let them go and thought, 'He's not bloody here.'

'Derbs' confirmed that the missing man hadn't entered the fort and so with that and the fact that there was a Ford back at the Assembly Area CSM Shepherd shouted, 'If he's not there when we get back we'll come straight across to find him.' And off 'Spinner' went with his troop.

'We now had no vehicles – there was me, the company clerk and I think the Fire Support Troop, which was about six lads, and the company commander's signaller and that was it – about ten or twelve of us marooned on the wrong side of the river and still being fired at. And firing back. I was talking to Captain Sharp, the FST commander, a fucking good man, a brave man, and told him that we had got to get our backs to the river and to do that we might have to cross on to the spit of sand.

'Captain Sharp asked, "Is there any cover there?"

'And I laughed as there was no cover where we were and anything was better than where we were – still being shot at from three sides. Any minute now we could be completely surrounded and I knew that if we reached the spit they could only be in front of us. So I said, "Stand by to move – get your kit together," and just as we were about to move, an empty vehicle suddenly appeared and with everyone in it was "toppers". I had one last look about, then ran and jumped in. The fucking thing was becoming a bullet magnet, so I shouted, "Go! Go! Go!" and slammed the door. We got through the river and across the other side where we sat on the ground and watched the fort being fucking malleted by the gunners, tanks and aircraft. When some explosives landed nearby I thought, "You fucking drop-shorts," as we call the Royal Artillery. But it wasn't ours, it was theirs, because we had moved far enough away for them to start mortaring us.'

Now that he could make himself heard, CSM Shepherd's concern – that there was a man still unaccounted for – was by no means over, so he called all the troop 'stripeys' to him and asked, 'Who have you got? I want another full headcount as the previous ones were taken in confusion.' It quickly became apparent that Lance Corporal Ford was not with them so, once more, the battle-hardened sergeant major spoke over the radio, now at last able to hear any responses clearly. This time he received the chilling reply: 'We have a Marine Ford at the Assembly Area – the CO's driver.' Shepherd's response was curt: 'Fuckin' 'ell.'

'When Zulu Company got back it filtered through to the HQ that a Marine Ford was missing,' Colonel Magowan explained. 'I said, "Calm down. Let's make sure who we have got." I was in my command post and a marine overhearing me said, "I'm Marine Ford," so I said, "OK, relax; fog of war, we have Marine Ford." Someone else piped up, "No, it's a Lance Corporal Ford that is missing."' To begin with, the colonel wasn't particularly worried: 'These things happen but then news came through from Zulu Coy that they had definitely left Lance Corporal Ford behind. Obviously very bad news indeed.'

To the west of the river and away from Jugroom, CSM Shepherd turned to the company commander and said, 'Ford's not here. He's over there somewhere. We have to go back in, we must look for him.' There was a stunned silence, broken only by the distant crump and thump of munitions falling on the fort. No one was prepared to say out loud what they all feared; not that Lance Corporal Ford was dead, but that he might have been captured – and that was frightening. What the Taliban had done to captured Russian soldiers during the Soviet invasion in the 1980s was well known.

CSM Shepherd was thinking rapidly. 'We can't just do this. We can't return in daylight, we can't just get in the vehicles and go over there as we have just had our arses kicked in the dark. We need to go, but we need to put this higher up the command chain as we need full support. We need a proper fire plan on to the positions.'

Despite the obvious reservations that they shared, his marines – and especially the whole of 5 Troop, who had been in the thick of it – were just as keen to recross the river. However, once the problem and the suggested solution had been passed up the line, the answer was, 'Wait: don't go over.' This came as some relief to the company sergeant major: 'I knew we were going to return but I wanted someone to sit and think this thing through calmly, so we were ordered to pull back away from the river to behind one of the fire support positions and into cover.'

From this new position Captain Witts, the company commander and CSM Shepherd were summoned to the LUP, about seven or so kilometres west into the desert. The Viking covered that distance quickly and within five or so minutes they were at IX Group HQ. 'Everyone who was anyone from the planning side of things was milling about until Colonel Magowan took us into a tent where just the battle group staff were.'

Witts, asked if he was happy to return, replied that he was: 'The colonel told us to have a cup of tea then come back with a plan. We did and our plan was to take four vehicles, surround Corporal Ford, stabilise him, then bring him home. There was of course one snag – we had no idea where he was.'

They left the commanding officer wishing them luck – while

others reminded Chris that he was due on his R-and-R in nine days – and returned to the river bank, where the volunteers selected by CSM Shepherd were getting ready, recharging their magazines. The plan was beefed up, the vehicles were mounted under Chris Witts's command, then driven to a position opposite the fort, where they halted for final clearance to proceed.

A sudden order – 'Stop! Stop!' – was shouted over the airwaves. The colonel had contacted the brigade commander at Lashkah Gah, on an indistinct line, to explain that he was missing one of his marines. The brigadier's direct response was 'to do everything that is necessary' to find him, and, in order to make that possible, went on to confirm that he was arranging for 'all the ISTAR that is available – Predator, more MR2, Apache, the lot – and then it will be up to you'. Zulu Company was champing to find 'Fordy' but as nobody had seen him, nobody had any clue where he was; the brigadier's idea to use all available air intelligence was vital in assisting them. By now it was broad daylight; nobody knew how many Taliban were left – although an estimate suggested around thirty – while the MR2 Nimrod had spotted reinforcements coming up from the Pakistan border.

At this point Sergeant Gary Stanton from the RAF – 'a very quiet and unassuming chap' – tapped the colonel on the shoulder and said, 'Sir, if you would like to come here a second.' Colonel Magowan followed him into the back of a vehicle to where the Desert Hawk's video was being beamed. The Desert Hawk is an unmanned aerial vehicle, like a model aeroplane, that is launched by three men using a huge rubber catapult; the engine starts when the aircraft reaches a velocity of 15m (50ft) per second. It supplied real-time video feeds, which Stanton, who'd been flying the Hawk over Jugroom Fort, deciphered now for the colonel as he pointed at his screen. 'That is a body and he is wearing a British uniform. I can't tell if he is alive or not but he is lying against the outside wall of the fort. Way to the left of the breach.'

A blurry image flickered on the screen, that showed the edge of the fort; on the left of the image a dark mark was clearly visible. It didn't move. Untrained as an image analyst, Colonel Magowan could only reply, 'What am I looking at? I can't see anything.'

'Believe me, sir,' replied Sergeant Stanton, 'that is a British body.'

The colonel immediately called up the brigadier. 'Sir, we've got him on video. I'm going to launch the lead company across the river with C Squadron in support and lots of Apache. I have no other choice unless special forces are available.'

But British Special Forces – preferably the Royal Marines's own Special Boat Service – were not available and, anyway, could not have been redeployed in time: although better trained in hostage rescue and similar situations they could have offered little more than that which was being prepared by IX's TAC HQ.

It was now about 0930. The Apaches that had been circling overhead and firing into the fort were aware of the problem. Warrant Officer First Class Mark Rutherford, a pilot with 9 Regiment Army Air Corps, called IX TAC HQ to ask for a Sitrep as his and three other aircraft were low on fuel. HQ's reply was that they were about to launch across the river and would need further air support; until then the Apaches were tasked to fly over Lance Corporal Ford's body on the ground, taking it in turns to loiter near the spot, guarding him. Rutherford replied that the plan would take hours, yet he could not supply top cover for much longer. 'I understand,' replied the colonel, unable to hide in his voice the feeling that without air cover, and in broad daylight, the troop going in would not have much of a chance.

The pilot knew this too, for he didn't take a moment to weigh up the obvious risks; the decision was clearly being made even as he spoke the words. 'Sir, I've a different idea. We will land at your location. Give me four men and I'll strap them to two Apaches.' Rutherford was making it up as he went along. 'We'll go over, land by the fort, strap the casualty to the base of my helicopter with a strop and come back. All you need to do is find four volunteers. We've never practised it – but...'

The colonel took no time to concur, and was straight on to brigade headquarters. This improbable plan was quickly agreed – although rumours later suggested that 9 Regiment's commanding officer had had serious reservations. Nevertheless the brigadier was happy for it to happen and, with that approval, responsibility was passed back down to the IX Group's TAC HQ. Colonel Magowan called the Apache pilot: 'Can you do it?'

'We're coming down now.' They were low on fuel and with no intermediate refuelling point between the LUP and Camp Bastion there was little time to waste.

The colonel had quickly to find the four men who would fly in on the Apaches. First to volunteer was Dave Rigg, the Royal Engineers officer who knew precisely what he would be letting himself in for, as all night he had watched the footage and, being the engineer, knew the ground. He knew, too, the strength of the enemy and had complete 'situational awareness'. Nevertheless, and without hesitation, he told the colonel that he wanted to go despite, in his CO's opinion, knowing that he had a greater than fifty/fifty chance of being killed.

The colonel accepted his offer immediately and looked around for three more. The Joint Terminal Attack commander volunteered but was vetoed: 'I need you to fire in the Apaches and B-1B bombers.' The operations officer, Major Sean Brady, volunteered but he, too, was vetoed: 'You are my ops officer.' At that moment the regimental sergeant major, Colin Hearn, poked his head round the tent flap. There was no time to go searching around the HQ: 'Ah, RSM. I've got a job for you.' RSM Hearn had been busy dealing with the wounded and was unaware of the situation. 'What's that, sir?'

'Get your kit together. You're getting on a helicopter to bring back Lance Corporal Ford.'

'This is a bite.' He clearly thought this was a wind-up.

'Afraid it's not,' came the reply. But the RSM was already preparing his kit; if it was a 'bite' he was going to be ready for it.

Another two 'volunteers' were still needed; the CO's signaller, hovering outside, having been relieved of his radio watch by the intelligence officer, was summoned: 'Marine Robertson, I've got a task for you.'

The reply was instantaneous: 'Very good, sir.'

Finally someone stopped a passing marine on his way to make a mug of tea. 'Fraser-Perry, the CO needs a volunteer.'

'I'll volunteer,' he answered, despite having no idea what for. 'Good, 'cos you're going on a helicopter, mate, back out to the fort.'

Having selected his 'volunteers', Colonel Magowan walked out to the helicopter landing site to greet the pilots. Dave Rigg was already there, wearing an expression described as one of 'I don't give a shit – I'm going to go and get this guy back.' The three marines, struggling into body armour and carrying their helmets, weapons and 'the normal' patrol kit, were running up to join him.

Dave Rigg had spent the night in the battle group command post; he had had access to all the imaging, had watched the Nimrod footage and had seen the hotspot that was Lance Corporal Ford, a fact now confirmed by the Apache pilots. As it was an hour and a half, maybe two hours, since he was likely to have been hit, Rigg thought 'Fordy' was probably dead, for he hadn't moved throughout the time that he had been under observation. The Apache pilots, on the other hand, with their thermal imaging equipment, believed him to be alive because of the warmth of his body; but there were others who considered that it was the rising sun that was keeping the lance corporal's body warm. No one was sure.

In the meantime, enemy reinforcements, still visible in the video feed from the Nimrod, were swarming up from the south, so with that and the possibility that the Taliban might at any time drag Corporal Ford back into the compound, it was clear to all that time was very short indeed. IX TAC HQ could have called on the B-1B bombers who had stayed on station to keep the Taliban at bay, but it was likely that any ordnance they dropped might hit Lance Corporal Ford; no one was prepared to take that chance. Added to this was the knowledge that the helicopters were perilously low on fuel; there wasn't time to think through what they were all about to do – they just had to get on and do it.

Dave Rigg hastily organised his team. 'Marine Gary Robertson was a stocky, tough-looking chap, so I said, "Right, you go with RSM Hearn." Chris Fraser-Perry was only about eighteen or nineteen with big, wide eyes and I told him, "You'd better come with me." The amazing thing was that once it became clear that they were about to undertake a pretty extraordinary bash, they didn't hesitate at all, and that was quite humbling. But I also felt a huge sense of responsibility because I didn't really know what I was leading them into either; I had to make it sound as though there was a plan but there wasn't – I still didn't know how we were going to hang on to those aircraft, or what we were going to do with the casualty once we had collected him.'

No one questioned the operation but, having readied themselves for a journey back to the fort, the men now began dropping their kit on to the desert around them, for they knew they would need minimal weight; the aim being speed above all else. They would take just their body armour, helmet and weapon. If they were to get bogged down in a firefight they'd lose the momentum of a surprise air assault, and at this stage the plan was no more sophisticated than that – fly in, jump off, grab their comrade and get him back to the helicopter. Two other Apaches flying alongside them would provide air cover to keep the Taliban at bay, so the men would not need their full kit. All the time they were preparing themselves, RSM Hearn kept repeating, 'It's a wind-up – a bite,' while all Dave Rigg could think was, 'Christ, is this what Bootnecks do to tease each other?'

As the men readied themselves, two Apaches swooped and settled behind the command post tents. Dave Rigg ran through the swirling dust kicked up by the heavy blades of the helicopter to the left-hand aircraft – thinking as he did so, 'Great, no briefing or anything' – and started shouting at Mark Rutherford, who was leaning out of his rear cockpit to listen. The pilot shouted a few words back but neither man could understand a word the other was saying above the roar of the engines and all they both achieved for their pains was a mouthful of desert. In the end Rutherford simply gave the thumbs up and screamed into Rigg's ear, 'Ready? Let's go!' He indicated that the Royal Engineers officer should sit on the rocket pod slung below the small stubby wing beneath the cockpit door, but Rigg shook his head and yelled back, 'Not yet.'

The mime show continued as the two men gestured at each other urgently, while failing to make themselves heard over the engine noise reverberating about them. Rigg's concern was that there was no obvious way to attach himself and Marine Fraser-Perry to the side of the chopper, but more importantly he needed to know how they planned to get Lance Corporal Ford back if, as expected, he needed to be carried, and who would do that while the others put down rounds to keep the Taliban away. Rutherford made himself understood, at last, when he handed Rigg a strop and waved it towards the front undercarriage, then pointed at the side of the fuselage where there was a fuel tank. 'It finally clicked that that is where we were to sit and that the strop was for Lance Corporal Ford and that we were just going to go in fast and pull out – there wasn't a lot more to it.'

But they still needed to run through a briefing. Rutherford climbed down from the Apache and the five men knelt in the sand. Rigg drew a quick sketch in the dirt, showing the positions of the fort, the breach, and where Lance Corporal Ford lay. This was vital as none of the others had seen the video feed he'd been watching, and if he didn't make it, then RSM Hearn or one of the others had to know where to go to in order to retrieve their fallen comrade. 'I pointed to the perimeter wall – marked in the sand – and to a position outside it and said, "That's Fordy; how close can we get to him?" then looked up at Rutherford. The Apache pilot studied the distance from Ford to the fort's wall, and calculated, "Within twenty metres," he offered.'

The colonel waved them off. Dave Rigg was happy now that a plan of sorts had been agreed but he was still questioning, as the men turned and ran back towards the Apaches, 'How the fuck is this going to work? We're going to have to sit somewhere – where? – and when we get there and find him, where do we put him, what do we hold on to?'

The waves of sand and dust whipped up by the helicopters engulfed them as they settled on to the Apaches. Dave Rigg sat on the starboard fuel tank, with the air intake banged up against his helmet, and with his right foot pressed down on the top of a Hellfire missile – 'I hoped to hell the pilot didn't need to press the trigger.' For some reason those on the left side of the machine were even more uncomfortable.

The blades rotated faster to take-off speed, the helicopter lunged upwards and forwards in the same movement; Dave Rigg felt his left hand involuntarily grip the strop he was holding on to even tighter. His stomach churned, not because he feared battle but because they were taking off into the unknown; nobody had ever tried this. 'Ridiculous,' he thought. 'We spend much of our time writing risk assessments and the next moment we're grasping the side of a helicopter gunship without a thought for our own 'ealth 'n' safety.'

As the Apaches rose into the sky the men clinging to their fuselages realised that the whole battle group HQ were standing outside their tents watching, taking photographs, and waving them off. Grim-faced, they realised that many of those waving were looking at them as if they thought they would never see them alive again.

The LUP was about seven kilometres to the west of the fort. The helicopters kept low to the ground and their speed down to fifty knots so that the marines wouldn't get blown off. The wind ripped into them, forcing dirt into their faces like specks of gravel, stinging bare flesh; thank God for their thick goggles. Dave Rigg was running through everything that he thought was going to happen in the next few minutes when he suddenly realised that in the four months he'd been in theatre he'd mostly been in a tent – and that he hadn't fired his weapon once. 'As I reckoned I would need to use it I decided to give it a little test fire, so fired at the ground – causing the front pilot to look round in shock.'

The desert was seldom empty, and Rigg had to be careful to avoid the shepherds and the camels; looking up from that rural scene, ahead under the bright blue sky, he saw for the first time the fort that up till then he'd only had the chance to see through the video feeds supplied by the planes flying overhead. The contrast, to him, was extraordinary; 'Beyond Jugroom, the sky was blue, goats grazed, families were going about their business, but the closer we got to the fort the more it resembled Armageddon: buildings smashed to pieces, imploded on themselves, trees and shrubs on fire, rubble everywhere, black smoke rising.'

As they drew closer to the target, the sound of gunfire could be heard even above the tumult of the helicopter's engines. For the two young marines clinging to the side of an attack helicopter, coming in fast and close to the ground, it must have been an horrific, eye-opening contrast with the comparative quiet and safety of the desert they had just left.

In the command post, Colonel Robert Magowan and his staff, watching the video feeds on their monitors, waited and held their breath. As the lead Apache approached the fort, they could clearly see the fading white patches where British artillery shells had exploded and, running back into the fort through the breached wall, little black spots – Taliban soldiers – returning to their fire positions. But then they were appalled to see the first helicopter flare out in an enveloping cloud of dust only to appear, seconds later, on the ground *within the fort*, firing its cannon at enemy positions as it landed.

From his precarious perch Dave Rigg was also watching, horrified. 'The first helicopter with RSM Hearn and Marine Robertson overshot the wall, firing cannon and rockets all the way into the compound – we weren't, as we were behind them, so couldn't fire for fear of hitting them – then landed, as he thought he was outside the fort, but he was concentrating on the enemy so much he completely misjudged the distances.' The passengers, not realizing where they were because of the sand blowing up all around them, jumped off to follow their orders – of laying down supporting fire to support the men jumping off the second aircraft. The two marines ran to the wall, expecting to be on the outside of the fort, but instead found themselves by an inner wall from which they came under intense fire: the Taliban could shoot through little holes, giving themselves excellent protection – but also restricting their arc of fire. So although it might have seemed the least safe place to be, inside the fort pressed up to the enemy's gun positions, the truth was that the closer to the wall they were, the safer it was for the two British marines.

Rutherford also lowered his Apache to the ground, but safely outside the fort, near to Lance Corporal Ford. Once he could jump down, Dave Rigg realised that he had a lot further to run to get to the stationary figure; he crossed the seventy metres as fast as he could but the ground was soft, rutted and 'hard as hell to run across', leaving the fit Rigg breathless. A building away from them erupted with small-arms fire and Mark Rutherford called upon one of the circling Apaches to hose it down with cannon, to protect the figure scurrying across the sand towards the wall. Watching on the video feeds the men at the LUP urged him on silently, waiting for the moment when Rigg would reach the motionless figure.

On the ground, though, it was obvious – immediately – that the Royal Marine was dead.

'Lance Corporal Ford was lying down; clearly he wasn't well but I couldn't see any blood,' said Dave Rigg. The officer knelt and tried to roll him over, but found he was too heavy to shift easily; he was a fourteen-stone man before taking into account the weight of his kit, armour, weapon and ammunition, which probably came to another three stone. Rigg could see, though, that Ford was very pale and it was clear that he had no pulse: 'I assumed he was dead and didn't take it beyond that but grabbed his webbing and started to drag him towards the helicopter.' At this point Marine Chris Fraser-Perry, who had jumped off the helicopter and run in the wrong direction initially, disoriented by the clouds of sand, caught up with Rigg. The two men tried to lift the body, worrying a little because they had been on the ground for some minutes, crouching beside Lance Corporal Ford long enough to become targets themselves. Dave Rigg believed that if he'd persevered, and with Fraser-Perry's help, he might have got Ford into a fireman's lift, but they did not have the time; and with the extra weight of his kit – as well as Rigg's own – along with the soft ground, he knew that they were not making enough headway back to the Apache.

'Fraser-Perry joined me and we tried to carry Ford back but were making a right balls of it. I wasn't thinking clearly as we were trying to lift him from each end with Fraser-Perry at his feet while I had his slippery upper torso. The process was not made easier by my weapon, which kept slipping off my shoulder, impeding the process, so I threw it to the ground and have received stick ever since. It's all the marines ever mention: "the chap who left his weapon behind".'

With Dave Rigg and Chris Fraser-Perry struggling outside the wall, inside the fort Staff Sergeant Armatage of the lead Apache had jumped out of his aircraft, caught up with RSM Hearn and Marine Robertson to shout that they were going the wrong way. The three of them managed to find an exit through the perimeter wall and ran over towards Rigg and Fraser-Perry.

By now Mark Rutherford had also jumped out of his aircraft and scrambled to join the others crouched down by the wall. He shouted, 'Don't worry about the gunfire, it's all ours.' Together, using his webbing, the three of them dragged Ford back to the helicopter 'where we put a strop round his torso, then attached it to the undercarriage bracing. As we were doing this the other three joined us.'

Hearn, Robertson and Armatage had run round the wall of the fort, being shot at as they went by the Taliban but also suffering the effects of the covering fire which the other two Apaches were laying down to protect them; empty shell cases tumbled around their heads. Exhausted by running through the soft sand, there wasn't much for them to do, so they turned round to stumble back across the uneven ground towards their helicopter. On the way they negotiated the bomb crater, so deep that it was 'like running into a trench'.

Rutherford's Apache, now with five men on board and short of fuel, flew directly to the desert LUP where there was great relief that Lance Corporal Ford's body had been recovered and that all the rescuers were returning unscathed. Everyone at IX TAC HQ was again outside, but this time silently waiting to welcome them home, relieved to see the Apaches return safely with their extra passenger, but deeply saddened by the sight of the dead lance corporal's body hanging below. 'It was particularly upsetting that we had to fly past the battle group with Lance Corporal Ford hanging limp beneath the aircraft, as we could see all his "oppos" looking up, praying that he was still alive.'

As soon as the helicopter landed, CSM Shepherd released his young lance corporal's body from the strop and laid him gently on the rough sand. The rest of the company stood at a dignified distance but Shepherd, believing that 'the lads shouldn't see this, he's dead', told them all to 'fuck off out of it'. They retreated a hundred yards. Zulu Company's sergeant major knew it looked callous at the time but he needed to recover some of the specialist kit off the dead body as, invariably and especially after a battle, there is a shortage of certain stores, items such as the individual night-sight attached to the helmet and other 'mission-essential' bits and pieces. These the CSM packed into an empty day-pack. Personal items were left in pockets, for they would be properly collected and tabulated at Camp Bastion, while Ford's rifle, helmet, webbing and body armour were stuffed into a second day-pack which was to remain with him.

Shepherd thought calmly, 'We are in the middle of the desert, we are not in a fight and we must get this right for young Fordy,' so he asked five or six of his closest colleagues from the same 'accommodation grot' at the Garmsir base to help prepare their friend for his last journey home. A lightweight stretcher and poncho were brought across and as he was laid out 'the lads were all talking to him quietly – it was rather emotional – as we wrapped him securely so that his arms wouldn't fall out while he was on the helicopter'.

The party then sat for some time, waiting for the Chinook to land and its medical response team to remove the body. That was how it was always done, but the men from Zulu Company wanted to be the ones to place him in the helicopter. It felt like a long time but it was probably no more than half an hour before the distinctive *whopper-whopper* beat of its huge twin blades could be heard, then swiftly the large helicopter was settling down nearby. They lifted the stretcher on board to find that the medical team hadn't been fully briefed, for one of them asked, 'What's wrong with him, where's he shot, where's he shot?' and CSM Shepherd thought, 'Fuckin' 'ell, why do you think we've covered his head over? And then I thought, they are doing their job as we do ours but it must have been flipping obvious.'

The men stood back from their comrade and said goodbye, one or two wiping tears from their eyes as they did so – then realised that the Chinook crew were looking at them strangely. They stood, bloodied, dirty, tired and looking for all the world like the ferocious warriors they had been a few short hours earlier; but now gently handling their friend's body and saying goodbye. 'We did seem a bit broken,' said CSM Shepherd. 'We were in a mess, we had just come out of a battle and we were delivering one of our own dead into someone else's hands. I thought, "Why are you all looking at us as though we're weirdos?" But I suppose the contrast between us and them was very real.'

The men who risked their own lives to retrieve their comrade felt they'd done only what any one of their fellow Royal Marines would have done for them. Chris Fraser-Perry who, only the day before, had been discussing mortgage options with Mathew Ford, said, 'I just wanted to reach Fordy as soon as possible and get him out.' Gary Robinson felt the same. RSM Colin Hearn said, 'I'm a Royal Marine and the RSM of the unit, he's a Royal Marine the same as me. There was no way we were going to leave him on that battlefield.'

With the Chinook departing through the familiar choking dust storm, Zulu Company's farewell party returned to their colleagues. In silence they mounted the Vikings to begin the drive back towards Garmsir, stopping at the same halfway position that they had occupied on the way out, even resting in the same shell-scrapes in the sand. The battle group's command post was collapsed and the BRF and C Squadron brought in.

Bombing, on its own, does not win wars, and whether or not raids by themselves help to win wars is another argument; but without sufficient men in Helmand Province no land, once taken by British forces, can be occupied for long, and the Taliban will be back. The country is simply too vast for every enemy position to be held permanently: that way lies stalemate, as continual enemy attacks against fixed locations had proved. Nevertheless, no enemy likes to be raided and he will spend precious resources in time, men and *materiel* trying to prevent such offensive – yet admittedly short-term – actions.

Everyone directly involved in the Jugroom Fort raid believed that it had achieved its objective, with IX Group's commanding officer voicing the majority opinion: 'This was a deliberate, pre-planned operation to disrupt the insurgents' freedom of movement in southern Helmand, a vital area from which they equipped and moved fighters into the centre of the province. Tremendous bravery, professionalism and endurance were evident across the battlefield by all troops involved in the operation.'

A major Taliban command and control centre had been destroyed and, in the words of a wise and experienced senior officer, the Information Exploitation Battle Group had done 'precisely what it says on the tin'. At Jugroom Fort, the brigadier's requirement for his troops to exercise 'dynamic unpredictability' had been well met: but at a cost. 45 Commando's commanding officer summed up their loss when he wrote, 'Lance Corporal Ford was a popular and gregarious young Royal Marine whose professionalism, reliability, and selflessness – as well as his sharp wit – marked him out from the crowd.'

After the extraordinary events of the day, Dave Rigg was left with one overriding impression: 'My own, humbling, memory is of those young lads prepared to give their absolute all for a fellow marine, and that will stick with me for the rest of my life.'

1
SEPTEMBER: OFFENSIVE SPIRIT

'As the Hercules was coming in I was amazed at the vastness of the desert and wondered how the Afghans could survive in such a biblical, barren land. It reminded me of the scenery in the Monty Python film The Life of Brian.*'*

3 Commando Brigade comprised 40, 42 and 45 Commando. For their posting to Afghanistan, 42 and 45 Commando were sent in September 2006; 40 Commando remained in Europe, training for future operations in Afghanistan. Arriving in country, the men of 42 Commando were rapidly dispersed to bases spread throughout southern Helmand and beyond, once they'd completed their Reception Staging Onward Integration (RSOI) training at Camp Bastion. Their commanding officer was Lieutenant Colonel Matt Holmes, a short, wiry, fit and vastly experienced officer who had fought in Afghanistan in 2001 and in every other operation with which his corps had been involved since joining in 1988, after reading economics at Exeter University. Qualified in jungle and arctic warfare, he had served in Brunei, completed three winters in north Norway and seen action in Northern Ireland, Iraq and Kosovo. Holmes was someone who, in short, knew what he wanted to do and how he would do it; he believed fervently in the 'offensive spirit'.

In this, as in so many other aspects, Holmes was superbly supported by his regimental sergeant major, a post initially held by Marc Wicks, a PT instructor, who handed over in late November to Si Brooks. Both were at the pinnacle of their distinguished careers, and both exercised a deep and caring concern for 'their lads'. It was rare for a handover to take place during a continuing operation, but the regular life of the corps had to proceed. RSM Brooks joined from 45 Commando; RSM Wicks moved to 1 Assault Group.

Between them the CO and the RSM were clear about the background to their deployment, as Holmes explains: 'As a battle group, our role was to enhance security to enable reconstruction and development to take place. We were there as part of a multinational force under NATO command and with the backing of the UN and the Afghan government. This differed from the Soviet invasion, as the vast majority of the population wanted us there. The Taliban on the other hand wanted to re-establish their rule which they could only pursue through violence and intimidation. In Helmand there was a distinct lack of governance anyway, so it was a natural harbour for the Taliban, not least because of their Pashtun links, the River Helmand that supports poppy growth, and the proximity of the international border. So we would have a fight on our hands in more ways than one, but as an internationally backed force we could offer so much more by working in conjunction with the Afghan government: the Afghan people just needed to be convinced and for that to happen there needed to be investment as well as security. This is not just a military campaign, far from it, which is why the FCO and DFID form part of the "Triumvirate" with the MoD, pursuing a comprehensive approach. There is little value in the military securing an area if nothing fills the void thereafter. Ultimately, it is investment in the form of reconstruction and development that will win the day, not courageous troops on the battlefield.'

Juliet, Kilo, Lima and Mike Companies were tasked by Lieutenant Colonel Holmes almost the moment they stepped off the plane. 'It was a very dynamic situation when we deployed; I didn't know which locations we would inherit from 3 Para, as there was some debate as to which platoon houses would endure. So I decided to deploy my companies in the chronological order they expected, so as not to mess around their families, with the proviso that they might go to a different platoon house on arrival.'

Travelling to their eventual operating bases in Helmand Province, the men were rapidly made aware of the conditions in which they would operate through the diminishing standard of the air transport that took them there. After the flight from the UK, cramped and basic Hercules took them the sixty minutes to Camp Bastion; then for those moving on from there to forward bases it was standing-room only inside the Chinooks – forty-four men with their bulging Bergens at their feet, crammed up against a quad-bike and trailer.

Deploying from the rear ramp of a Chinook in Afghanistan for the first time came as a surprise to many of the marines on board; the dust, whipped up in the downdraft, enveloped the disembark-ing men in thick choking clouds, sometimes hurled about with such ferocity that breathing in as they stepped off the aircraft could provoke nosebleeds.

There was usually a welcoming party – the Taliban made sure of that – so Chinook pilots were always keen to touch down and be gone as fast as possible to avoid any chance of being shot by the enemy. Sometimes they'd be so anxious to get away that the men would have to jump out of the helicopter on to the rough stony ground, scraping hands and jarring knees on impact, cursing as they did so. A few bullets and perhaps an RPG or two popped over their heads would signal the Taliban were out there, watching, as the marines cleared the landing zone as fast as they could.

To help protect the Chinooks, the landing site at Sangin was shielded from direct attack by a perimeter of collapsible Hesco Bastion containers filled with rubble that provided good protection from small-arms fire and RPGs. As the helicopters circled round, readying themselves for landing, every Lima Company marine craned to peer through the Chinook's dusty windows at the danger-ous place they had heard so much about. Marine Hayward later jotted down his feelings: 'Apprehension and excitement as we dropped into the unknown. Couldn't see five feet in front of me – trying to stay on my feet and find my section commander. Couldn't believe how different the reality was from the picture in my head.'

Barren was the word for this desolate landscape; spectacular from above, it was brown, gritty and monotonous once the marines were on the ground. The air was dry and filled with the smell of animal dung; in the midday heat, a breeze would pick up and flap away at tents, or clothing, or anything not pegged down. The heat was so intense in the middle of the day that it hit the back of your throat the moment you opened your mouth. What at first seemed remark-ably different to marines like Hayward – apart from the feeling that they'd seen this place before, but only in a Monty Python film – was the lack of action. For a number of Lima Company's marines, many of whom had not long been out of training, the adrenaline rush of arrival was short-lived for, by coincidence or not – and nobody knows – the Taliban declared an unofficial ceasefire. Colonel Holmes was puzzled. 'Having put the company in there expecting a very kinetic, high-risk period – there was nothing. I really don't know why; but it was agreed between the Taliban and the governor and on that basis we were constrained to the District Centre and not able to patrol out. It was very frustrating for the marines who now had to maintain their discipline and keep their guard up; never knowing if or when the cessation of hostilities might break – which it was to do in March 2007 with a wave of concerted attacks against Mike Company during their first day in the District Centre.'

Lima Company, 42 Commando, under the command of Major Liddle, had arrived at Camp Bastion on 22 September, and went through a period of RSOI training, a package designed to prepare them – and every other new arrival – for what was about to come.

In effect it was a useful few days based on the experiences of those departing. Steve Liddle had been commissioned into the Royal Marines in September 1990 from Oxford University, later joining 42 Commando straight from the Defence Academy. He had recently celebrated a second year in command of Lima Company with his commanding officer considering him 'intelligent, level-headed and ready for a challenge'.

Matt Holmes had moved Lima Company into what he thought would be the thickest of the action because of the men in the company. 'I put them into Sangin because at the time it was the centre of the hardest fighting and Steve's sergeant major was Billy Bowker, a hard fighting man, exactly the sort of company sergeant major I needed in a place like that.'

Sangin was expected to be a tough posting for the marines because of the decision of the Paras to base themselves inside fortified buildings, then wait for the Taliban to come and try and prise them out. Colonel Holmes – among many – had never been convinced of the Platoon House strategy, 'The CO of 3 Para claimed the Platoon Houses enabled the destruction of hundreds of enemy who were drawn to them. They were originally manned at a request from the Governor of Helmand, who wanted to prevent them falling into Taliban hands; the local theory being that if you held the bullseye you controlled the dartboard. The destruction of swathes of Taliban was never in my opinion the reason for manning these isolated locations. They were certainly dispersed, and came under constant Taliban attack. On the basis of the ink-spot or oil-slick theory, that influence can extend from these centres, we may just have flicked our ink pen a little too hard at the start of the campaign, but therein lies some of the complexity of the situation, as the governor needed a presence in certain areas. Our view on deployment was that the Afghan Development Zone, not specifically defined in geographical terms but centred on Lashkar Gah, the provincial capital, required urgent attention and investment and was the logical place to start making a difference.'

In the initial plan for this whole operation, drawn up by Colonel Gordon Messenger of the Royal Marines when he had deployed with Joint Force Headquarters, the focus was to be on the Afghan Development Zone, centred on Lashkar Gah. 'The Taliban want to rule Afghanistan again so, locally, they had to defeat us. I suspect, though, that this aim of theirs may have been reviewed: inflicting casualties and discrediting us is probably higher on their list now.'

With the Royal Marines now in control after an all-night handover, the Paras – bearded, tired and relieved – were at last, after ninety-one days under near-constant attack, able to board the Chinooks and head for Cyprus, home and a well-deserved leave. All, that is, but for one Parachute Regiment corporal trying to load his quad bike on board the last Chinook. He was waved away by the RAF loadmaster as the aircraft began to lift, the reluctance of the crew to linger on the ground any longer than necessary preventing him from getting the Chinook's payload rearranged to accommodate the company's bike. The corporal had to wait another four days before he could start his journey home.

While the Platoon House strategy may have added to the attrition of Taliban fighters, the commando brigade's view was that a simple head count was not the aim of the operation: 'The destruction of people and property risked alienating the very citizens we were trying to support and help, while accepting that the *irreconcilable* Taliban would have to be isolated for destruction.'

Of course the Royal Marines were fortunate as the manoeuvrability aspect, not available to their predecessors, was given a huge boost by the arrival of the Viking all-terrain, protected vehicle. This is not an armoured fighting vehicle, although it was often to be used as one. How differently the Parachute Regiment would have conducted their affairs had they had Vikings and more men can only be conjectured. The Vikings, though, had yet to arrive in the country, for they were still en route through Pakistan.

Ceasefires, Helmand-style, do not imply peace: throughout their time in Sangin shots were fired against Lima Company almost daily but never returned. Their chance for more direct action would come at Now Zad a little later in the tour, but in the meantime, and for nigh on two months, they kept themselves fit in a mini-gym they constructed themselves in a corner of the disused police station, and watched and waited. The most significant threat to the company's welfare over the first few days was not indirect RPG fire but diarrhoea. No matter what precautions were taken, an insidious bug took its toll, just as it did in all the District Centres and Forward Operating Bases. A public line of oil drums stood in place of flushing lavatories – or 'heads' as the marines call them, reflecting their ship-borne history – with the contents burned off each day. When possible, the marines would use makeshift wooden seats perched over a pit and made 'private' by blankets strung around at shoulder height.

Conditions were spartan; fresh water was at a premium and lighting at night was by candle apart from single lights in the medical aid post and galley.

Setting out on patrol was governed by the thought that the men could not be sure when they would return, so they had to be prepared for any and every eventuality. The bulk of kit carried by each man varied from task to task, mission by mission; most of the equipment consisted of ammunition and water. To these would be added mission-essential equipment, depending upon the type of patrol. Water was vital; for most patrols, whether dominating the area or for 'hearts and minds', each man would carry at least three litres of water in 'camel-back' containers.

Commanders would carry their own personal Bowman radios, plus two spare batteries and PRR – personal role radio. Each marine had sixteen magazines (twelve in pouches and four in a day-sack), three or four red phosphorus grenades, the same number of HE – high explosive – grenades, mini-flares, 51mm mortar bombs, GPMG ammunition, night-vision equipment, spare batteries, maps, GPS, an air-panel marker, mine kit, a very small amount of food – either boil-in-the-bag or biscuits – and finally, the Osprey body armour that, with both protection plates in place, weighed heavily.

Every man had a specific role: a patrol 'medic' would carry appropriate equipment for that task, as well as extra section- or troop-stores; each marine would carry various types and scale of ammunition depending on their weapon type; a GPMG gunner would carry around 1,000–1,500 rounds of 7.62mm link in addition to much of the above, while Minimi and grenade-launcher gunners carried extra bombs and ammunition. Spread across each section were two large items of ECM equipment loaded in a day-pack and also weighing heavily.

For a deliberate assault task, extra ammunition and explosives for mouse-hole entries would be carried, as well as AT4 rocket launchers to engage dug-in or defended enemy positions. On almost every patrol the equipment was heavy, with the only sacrificial commodity being food; weighing up the options of either hitting hard and moving fast, or being patrol-heavy for sustained firefights was always a dilemma. If mounted as the Fire Support Group or a Quick Reaction Force, kit could be loaded on the vehicles. Such equipment might include Javelin anti-tank missiles with their CLU (command launch unit), GPMGs and .5-calibre heavy machine guns. On occasion, for deliberate assaults, the FSG would lift all the vehicle-mounted equipment plus ammunition and weapons to the fire support area.

Patrols could vary in time and length; what might be planned as a routine three- to five-hour patrol in the local area could easily turn into a thirteen-hour patrol in contact, offensive and defensive. Some local patrols could be extended to twenty-four or even forty-eight hours: the most common were eight- to ten-hour patrols, both by day and by night. As temperatures dropped in the dark and in winter, cold-weather layers had to be worn, again bogging the individual down. Inevitably, on occasions no warm clothing would be carried.

Juliet, Lima, Kilo and Mike all staged through Camp Bastion: a sprawl of tents, hard-walled buildings and stores dumps that lies in open desert to the west of Gereshk and south of Highway One, a hard-topped road that bisects Helmand Province as it runs roughly west to east. Built in early 2006 by British Royal Engineers, the camp was designed to support British forces in the province and, measuring two miles by four, is the largest British military base to have been built since the Second World War. Juliet Company had been there before, guarding the Royal Engineers while they constructed the camp.

Captain Luke Kenny, of Kilo Company, under the command of Major Neil Sutherland, described their reception at Camp Bastion on 23 September. 'We arrived in the afternoon, giving us time to settle in and unpack our kit before the RSOI started the following morning. The air conditioning was broken in the RSOI tent but the guys remained focused throughout the whole day in unpleasant conditions, knowing that all the points made were pertinent to them being out there and thus to their basic survival. We were given a full day's briefings on intelligence updates, environmental health, spiders and snakes, driving, first aid, camp routines and the welfare packages available.'

The following day the marines of Kilo Company were out on the ranges, 'prepping' and testing all their weapons, at the same time learning how to deal with the dust that rose and settled into every available nook and cranny. It was bad enough with dust in their ears and eyes, but worse when it jammed their weapons; now they had to relearn how to clear and clean their weapons while lying prone in the dirt.

Commando training ensures that within the Royal Marines there are numerous outstanding young leaders, many without stripes on their arms or pips on their shoulders, so it was the brigadier's express wish, now that his marines were in country, that the 'offensive spirit' be expressed by leaders at every level of the command chain. One such man – although in this case an experienced corporal and section commander – is Al Hewett, who was to win the Military Cross a few weeks later leading his section through a maze of compounds north of the Kajaki dam. '1 section, Kilo Company, formed in late 2006 during pre-operational deployment training for the forthcoming Afghanistan tour. I knew that the training of my men was imperative for the high-tempo fighting that they were about to experience. I first met my section on the Otterburn training ranges and was taken aback by how young and relatively inexperienced the lads looked. First appearances are not always the correct ones and to the good fortune of us all we formed a very strong section. This, in no uncertain terms, was down to the superb efforts of Lance Corporal Tom Birch, my second-in-command.'

With their RSOI completed, Kilo Company deployed into Forward Operating Base Price a kilometre or so to the west of Gereshk, to conduct a Relief in Place with a company of 3 Para. This was a temporary, if dramatic, assignment while they held the fort, literally, for ten days, until relieved as planned by Major Ewen Murchison's Juliet Company.

Colonel Holmes explains. 'FOB Price was a mini-Camp Bastion that we shared with an American Task Force, Operational Deployment Alpha, who worked fairly randomly with Afghan militias but without co-ordination with us. They just went off to kill Taliban with little thought of the wider picture, presenting us with significant risks of blue-on-blue engagements.

'Gareshk had been neglected and yet it is the north-east gateway to the Afghan Development Zone, which extended south-west and obviously included Lashkar Gah, the provincial capital. As it straddles Highway One and the Helmand River its strategic importance is clear especially as, to the north-east, a substantial irrigation canal leads from the Helmand River via a sluice gate some six kilometres away; this waterway is dammed about two kilometres from the town where a small hydroelectric plant supplies power to the town. The whole complex was of considerable interest to the Taliban who would come down the river and the canal to attack the Afghan police guarding the dam: embarrassing the authorities. Gereshk is also the gateway to Lashkar Gah, further south, so it had to be my main effort. As Ewen was my senior company commander I gave him responsibility for this independent and vital location. The canal was not fordable while the river was, but only in the summer. The three bridges meant that one could drive into Gareshk and not need air. Everyone knew that if the Taliban controlled the bridges they would have complete freedom of movement along the road and be able to operate with impunity.'

Neil Sutherland of Kilo Company describes the first busy days: 'We deployed to Gereshk in order to fill the gap behind a Para company who were scheduled to leave before Juliet Company arrived. The CO had in his mind who he wanted where and while Juliet completed their RSOI at Bastion it was us on a temporary basis.

'We became quite familiar with the town. Being on the road and river, it is in a key position and differs significantly from some of the more northern and eastern bases such as Sangin, Musa Qal'eh, Kajaki and Now Zad. It is far more built-up and quite large; some estimates put the population at about thirty thousand, with FOB Price a defended compound a few kilometres to the west.'

Gereshk, one of the wealthier towns that border the Sangin Valley, is basic by Western standards with local townsfolk indifferent to the Royal Marines' arrival. Nevertheless, through a positive presence on the ground, Kilo and then Juliet Company began to win over the inhabitants and to enjoy an element of freedom of movement throughout the bazaar and the surrounding area. Beyond the bazaar, though, the welcome was always less warm and, particularly to the east, downright hostile and dangerous, for these villages were home to Taliban fighters.

From FOB Price's well-defended location, the men of Kilo Company began their patrols in and around the town, sometimes on foot but always having left their base in vehicles. There were two aims to this programme. First, it was good ground familiarization so that the men could pass on useful intelligence to their successors and thus save them time; second it was an opportunity to get the officers, NCOs and marines used to operating in that environment and among the Afghan people. Neil Sutherland also needed to settle them down before moving on to potentially more hostile locations at Kajaki and Now Zad.

Representatives of the local police force came to meet Sutherland and his men the day after their arrival at FOB Price. They brought with them news that the Taliban, having taken note of the change of personnel and, presumably keen to test the mettle of the new arrivals, were again active. So the following day Kilo Company set out determined to make sure that their presence was indeed noted by all those who needed to see them, while also familiarising themselves with the area to the south of the town, east to the canal and then north-east towards the hydroelectric power station and the dam. Additionally, the patrol was tasked to identify various places where permanent checkpoints could be established, to protect the routes in and out of town. More importantly, they needed to work out where a checkpoint could be placed that would cover the dam itself, near to the position where the canal connected with the Helmand River.

The patrol deployed in a mixture of vehicles: Land Rovers with mounted weapons, known as WMIKs, and Pinzgauers, open troop carriers. At a pre-set drop-off point on the outskirts of the town, the men dismounted, 5 Troop leading and company HQ behind with the Fire Support Group in a quick reaction force role. 'The area was close country, right on the edge of the green zone, so we left our vehicles about one kilometre back.'

The patrol, with Corporal Al Hewett's section to the fore, made it to the dam without the enemy emerging to confront them; from there the route swung north towards higher ground, where they would link up with an ANP station. As they came closer to the checkpoint a policeman emerged, waving frantically at them and pointing towards the valley. It was clear that the Taliban had moved into a position from where they were preparing an attack.

'The recce on the position was short for we were by now exposed in open ground, but as this was the end of the outward-bound leg we turned back towards the pick-up point with my section now in the rear,' said Corporal Hewett. It was clear from the enemy radio chatter, and the rate at which it was increasing, that the contact was imminent. The troop moved up on to higher ground, overlooking the dam and the lower valley area, some 600 metres from the canal. As the patrol moved out of sight of the dam and among the compounds, a large explosion came from the direction of the canal. 'I looked at Sergeant "Penny" Farthing – our troop sergeant – who agreed it must have been in the dam area. At the same time the troop boss – Captain Luke Kenny – asked me over the radio to continue moving towards the high ground and the police checkpoint.'

Now the marines themselves came under hostile fire. Initially it seemed little more than a few men taking potshots at them, but as the section took cover more of the Taliban joined in the attack with mortars now falling 'like rain' among the marines. It was also the first time they had encountered the 'Afghan bees' – the bullets that zipped and whistled past their heads – as well as the thin *whoosh* of the shoulder-held RPGs.

'I positioned the lads against the wall of a bombed-out compound that overlooked the dam, some twenty metres from the

ANP checkpoint, then ordered Tom and Gav Wignal around the side of the mud wall from where they could see the site of the explosion. After another burst of machine-gun fire past our cover I ran over and linked up with the police guard on top of the hill, at his outpost.'

From his position overlooking the dam, Hewett was able to see the layout below, watching as the Taliban advanced. 'I could see a group of police guards in the low ground some 200 metres in front of me who were being engaged by RPG and machine-gun fire. The Taliban were moving up to try and capture the dam with the police now falling back under fire.' As he marked the positions of the enemy, the shriek and smoke indicated another large rocket, heading towards the dam. The ground erupted again from another blast and this time Hewett was able to confirm to his section that it was a 107mm rocket.

Although the men were well positioned, on the high ground and on the same side of the river as the Afghan National Police, Neil Sutherland knew that they had 'to win the firefight to push the Taliban off the dam'. Having left their vehicles behind them, they weren't going to benefit from their heavier machine guns, but not only did the Royal Marines have their own mortars with them with which they could peg the Taliban back, they also had access to something the Taliban did not: aircraft armed for air-to-ground strikes.

Corporal Hewett moved his LMG – light machine gun – gunners, Matt Fenton and Dave White, further up the slope to better target the Taliban moving on to the dam. Using the LMGs and 1 and 2 Sections, Hewett aimed to keep the Taliban off the dam long enough for the air strike to be called. 'Fast air', as the Harriers and other attack aircraft were nicknamed, would neutralise the Taliban promptly but not if they had already taken up positions on the dam itself. They couldn't risk damaging that.

For another half an hour, maybe a touch longer, the British kept the Taliban down the valley, ensuring that the enemy took cover, thus preventing them advancing on the ANP positions, and thus away from the dam. It was an intense, small-arms firefight, with the heavier weapons back on the vehicles they'd had to leave behind. Eventually, however, a distant but increasing buzz indicated that the tide was about to turn firmly in their favour. Apache attack helicopters and a Harrier GR7 were moving in and even at a distance were already able to target the enemy forces, with the company's Fire Support Team passing grid references to the pilots. As the Taliban retreated – still firing – into an area free of civilians, a Harrier launched two salvos, twenty-six rockets in all.

'And that was the end of that! There was no further sign of the Taliban. They had either been killed or had managed to melt away, which they were very good at doing,' said Sutherland. This was the first time he had called in air support, and he and his men were heartened by the result. It was also clear that the Taliban did not like the fast-moving, highly reactive Apaches.

Sutherland – in company with a number of his fellow Royal Marines – thought the Harriers 'fantastic'. 'We were confident that once we had a GR7 overhead the pilot would be able to identify even the most difficult of targets and would do precisely as asked. Whether or not it was flown by a Royal Navy, Royal Marine or RAF pilot we didn't mind – it was a Brit!'

But in the coming months it was the Apaches that most endeared themselves to the marines. The UK Apache Longbow helicopters, unlike their US or European counterparts, are fitted with Rolls-Royce Turbomeca RTM322 engines, which allows them to accept a heavier payload: thus they can lift their fire-control radar systems into battle. As a result the pilots not only have immediate awareness of the position of other rotary-wing aircraft in their vicinity, such as Chinooks in poor visibility, but can also 'see' the ground – known as 'terrain profiling' – which helps to identify dips and hollows in the ground where the enemy could be hiding. More importantly, the FCRs – the thick black 'bulb' on top of the helicopter's rotors

– can register unknown vehicles and identify friendly ones, using the tagging system that allows the pilots to 'see' items as small as a medium-calibre machine gun.

Once a target is acquired, the Apache bristles with the weapons to destroy it. Apart from the 30mm cannon, with a range of over 4,000m and a firing rate of 600 rounds per minute, the helicopter carries semi-active, laser-guided Hellfire missiles. Because of their substantial anti-armour warheads, these can be used to blast open walls behind which the Taliban might be sheltering. In addition, the Apache can carry CRV7 70mm rockets with an astonishing range – 8.2km – and variable warheads: flechette, semi-armour piercing, or penetrator types. With the four launchers full, the Apache carries seventy-six of these rockets.

Finally, because of its manoeuvrability and the precision of its weaponry, the Apache – unlike 'fast air', which requires specific pre-planned targets – can act in close support with ground troops and, using its cannon, can even fire on the enemy as close as ten metres from 'friendly forces'. This is known as 'danger-close', with the Royal Marines able to feel the effects – and be grateful for them – on many occasions throughout the months ahead.

Little wonder, then, that the Taliban feared the Apache.

Corporal Hewett stayed at the ANP station with his section for some time, waiting for any sign that the enemy was recovering from the aerial assault. 'After an hour or so in position watching for a possible enemy counter-attack, the troop and company HQ moved to the pick-up point and back to FOB Price.' So ended Kilo's first contact – a useful patrol from the company commander's perspective: 'My marines had successfully seen off an attack, there were no casualties and the lads had been under fire for the first time in Afghanistan, so that was a box they could tick.'

Sutherland had been concerned – a concern felt by all commanders during this introductory period – that days of waiting for a contact and thus waiting to answer the unasked question of how an individual would behave 'in contact', would prey on minds. Being shot and mortared is never comfortable, but each individual's worry could now be slightly assuaged. 'We had supported the ANP, we had not killed any civilians and had been robust in our response. The fact that we had also killed many Taliban was good for the lads and a good bedding-in experience for the company.'

The third of 42 Commando's companies, Juliet, commanded by Major Ewen Murchison, a bald-headed, hefty man and a former Scotland under-21 rugby international, was tasked by Matt Holmes to be the permanent patrols company based at FOB Price. The colonel's overriding plan was for Juliet Company to establish and strengthen the Afghan Development Zone in order that reconstruction and firm, legal governance could be expanded.

So, after just ten days in Gereshk, Kilo was relieved by Juliet, whose men were once more back in Helmand Province; their first visit to what was then a flat barren landscape had been in February, when they protected the Royal Engineers constructing Camp Bastion. Now a small, military city had sprung up from the Afghan sand.

Fire Base Price or Forward Operating Base Price – depending to whom one was talking, US or British – already housed the American Deployment Group Alpha who, as Murchison described, were 'doing normal special forces stuff with surrogate-type forces: ANA and those sort of people. So we were in this small American camp and like all these things there may have been restrictions at more senior level but at local level there was no problem: we scratched each other's backs in different ways. We had to live with them whether we or they liked it or not – and it was their camp anyway.'

Colonel Holmes made it clear to his senior company commander precisely what it was he wanted from Juliet Company: 'This may not seem like the sexiest task at the outset and I know you and your men want to be out there where the action is. I need you to go into Gereshk and work up the relationships with the locals in order to develop the Afghan Development Zone and the Afghan Security Forces. Then, let's see how things develop from there.'

As it was to turn out, Ewen Murchison and his men were, arguably, to have the most interesting tour of all 42 Commando's rifle companies, as one of his marines explains.

Marine Mark Farr had joined the corps late, having graduated with a degree in German and Russian from Bangor University and, hardly surprisingly, had been appointed Juliet Company's intelligence representative: for which he had attended a two-week Tactical Intelligence Course in June, learning the basics of collation, interpretation and dissemination.

'Our tour began in Gereshk, where we were fortunate enough to stay for the entire time. This saved considerable upheaval and admin nightmares. I went out at the beginning of September, some three weeks before the company, to get a decent handover from the Paras, a feel for the area and to prepare a brief for Major Murchison. I joined Alpha Company of 3 Para under Major Jamie Loden; a very good bunch of lads who had both seen and done a lot. As a Bootneck, the name by which we marines are known, they made me feel more than welcome and even respected me for being one. I met a number of lads I had seen when they took over from us in April 2006. One of the Paras came up to me and said, "Do you recognise me? Six months ago we were chatting and you told me that you only had ten days to go before you went home. Well, now it's my turn, I go home in ten!" Down to a man they were a good bunch, pretty honking, as all Paras are, but a good bunch none the less.'

As Juliet Company's intelligence representative, Farr was privy to the opinions and views of the command as well as those of his peers. 'When it was discovered that we were going to Gereshk there was a lot of piss-taking banter from the other companies, as until then Gereshk had been seen as a bit of a backwater; while the Paras had been there nothing had really happened.'

Central to Colonel Holmes's aim was the development of the Afghan National Police, although they were not a police force as those in the UK would know one. Alongside them were the Afghan Highway Police, an even more doubtful organization of semi-uniformed miscreants, who extorted money from civilians using Highway One. While the Afghan National Army were recruited on a national level, the Highway Police were recruited locally; thus bribery and local loyalties reduced their effectiveness. The Royal Marines Police were tasked with training these men, on the assumption that the sooner the Afghans controlled the town themselves, the sooner it would free British troops to concentrate on redevelopment and defeating the Taliban militarily. The Royal Marines provided them with uniforms but the police tended to wear them only 'when they knew we were about to visit'.

Supporting Major Murchison was his second-in-command, Captain Bruce Anderson, also from Scotland, and three troop commanders, Chris Burr, Gordon Sweeney and Tom Scott. At the core of any company is, of course, the sergeant major, and Juliet Company had Scott Burney; much liked by the men as an efficient and caring warrant officer, able to maintain a balance of both good humour and discipline. A platoon weapons instructor by trade, he had been the CSM when Juliet protected the building of Camp Bastion between February and April 2006. It was the custom to see him, first thing in the morning, clutching a cup of filter coffee, something he loved almost as much as his cigars.

The final unit of 42 Commando was Mike Company, commanded by Major Martin Collin, a jungle warfare instructor who, among many other operational theatres, had seen service in Sierra Leone, where he had trained the local Force Reconnaissance Unit. His second-in-command, Captain Anthony Forshaw, who was to lead the company in a set-piece company attack while Martin Collin was on leave, had set his heart on join-ing the Royal Marines. After university he was told he was too old, so, knowing that the regiment had an exchange posting with the corps, he was commissioned into the Argyll and Sutherland Highlanders from Sandhurst and was then in a position to apply for – and be granted – that exchange. Shortly after joining Mike Company he asked for a permanent attachment to the Royal Marines and was fortunate that his transfer came through on 1 April 2006; he thus arrived in Afghanistan 'as a Royal Marines captain, and was pretty chuffed at that'.

'Our start was frustrating as "Ops One Company" retained at Camp Bastion and ready for use in independent operations or if someone was in the shit and needed pulling out. I had one troop at thirty minutes' notice to move,' said Major Collin, 'but the whole company was, in effect, at the same notice and certainly no more than an hour. In two months we only had three small actions.' Nevertheless a small portion of Mike Company was involved in operations earlier than the main body, for 10 Troop, commanded by Captain William Mackenzie Green, a mixed troop of general-duty riflemen and support weapons operators, was despatched on 25 September 2006 to overlook Kajaki dam from two hilltop observation posts (OPs). The turn of the rest of the company would come later – in spades – at Kajaki and Sangin.

The OPs at Kajaki lie on the four peaks of a two-kilometre-long ridge some 500 metres above the river, to the south and west of the actual dam. The OPs were named – from the east – Athens, Normandy and, after they were captured, the twin peaks of Sparrowhawk East and Sparrowhawk West.

When 10 Troop relieved the Paras, Athens was occupied by a mortar pit. The 'Toms' had been occupying the two eastern OPs since the summer, using their 'tubes' and 'fast air' to engage anything they identified to the north. Sparrowhawk East and West were a 'bit of a battleground between the Taliban and the local militia' who looked after the bridge that crosses the river about one kilometre below – to the west of – the dam; the militia had also occupied three high points on the north side of the river below the dam and just to the south-east of Qal'eh-ye Gol. These latter three posts, forming a triangle, were given the uncharismatic names of Unknown Left, Unknown Right and OP 13. Further east, almost overlooking the dam complex and also manned by militia, was OP 14; these militia were paid by the Americans, whereas the others were paid by the Afghan police.

The militia and the Taliban had taken it in turns to occupy Sparrowhawk East and West; one side would take the occasional trip up, get shot at, and then it would be the turn of the other side. The Paras had made life difficult for the enemy by opening up on them whenever they saw their 'soldiers' on the ridgeline, but they never went after them. Out in the open and off the well-trodden tracks, it wasn't safe, even when the Taliban were beaten back, for the hillsides were covered with mines, often dating to the war with the Russians in the early 1980s. In early September 2006, a party of Paras had set out to conduct a couple of satellite patrols around Normandy OP, but had quickly found themselves stuck in a minefield.

With those experiences at the back of their minds, it was no surprise that Will Mackenzie-Green's troop of twenty-eight marines sat tight in the OPs on Athens and Normandy for the month they were posted there. Like the Paras, there was not the manpower to take the fight to the Taliban; if they'd come up against any serious force they would have had little or no chance of extracting any wounded, for they had just one Land Rover. Sitting tight was the best answer, as they had two mortars and a handful of heavy weapons: a .5-calibre (.5-inch) belt-fed heavy machine gun and – useful at the distances they were from Sparrowhawk East and Sparrowhawk West – Javelin, the lightweight, portable, anti-armour, fire-and-forget weapon used in a 'bunker-busting' role, called 'fire-and-forget' because the missile is guided by an infra-red, lock-on-before-firing system. With a range between 65 and 2,500 metres, it is an excellent weapon for long-distance attacks against static targets.

The marines of 10 Troop were there for about a month before they were 'Relieved in Place' by 5 Troop, Kilo Company; they, in turn, handed over to the whole of Mike Company, some months later, deployed to conduct a wider range of offensive operations.

For the majority of Mike Company, though, it remained a frustrating period as Camp Bastion's Quick Reaction Force for, as Martin Collin said, they were deployed on only three occasions: once to clear an improvised explosive device (IED) and, on two separate occasions, to protect the men of 28 Engineer Regiment who were constructing permanent vehicle checkpoints in the Gereshk area.

While the Royal Marines were taking over from the Parachute Regiment in Helmand, staff from 3 Commando Brigade flew to Kandahar Province to begin their takeover from 16 Air Assault Brigade, under whose command 3 Para had had such an eventful tour. 16 Air Assault Brigade had established their HQ in Kandahar, although the Paras themselves had been based in Helmand, which the Royal Marines thought to be 'incredibly messy'. Jerry Thomas believed that if he were to command the Helmand Task Force he had to do so in Helmand. There were arguments against this; for instance the helicopter landing site, being inside the base, would bring an increase in aircraft movement; but the real threat in Lashkar Gah came from suicide bombers, about one every six weeks.

The commando brigade's chief of staff was ordered to establish the brigade HQ in Lashkar Gah by 6 November: a quick call bearing in mind that the commando brigade officially took over from 16 AAB on 9 October. Nevertheless, everyone was moved from a comfortable hangar in Kandahar to 'gash tentage' in Lashkar Gah from where the advance party was able to gain valuable 'situational awareness' early on while it waited for the brigadier.

By placing his headquarters forward, Brigadier Thomas demonstrated to the other government departments in the area that he was part of their set-up, arguing that one of the many reasons he had to be in Lashkar Gah was that that was the site of the Provincial Reconstruction Team and that his brigade was there primarily to aid and enable that process.

As Jerry Thomas explained to the senior British government representative in Helmand, first Nick Kay and then David Slim, 'In the Afghan Development Zone you are the *supported* commander with me s*upporting* you for reconstruction and government, whereas outside the ADZ I am the *supported* commander because I need you to think about development and government once I have established contagious security.'

David Slim, who had been Ambassador in North Korea, was considered 'a brilliant guy' who 'understood what was required and was desperately trying to push the boundaries but within his own constraints'. As a staff officer was to comment, 'The Department for International Development is not an organization that contracts things; if you don't have any aid agencies in Lashkar Gah you don't have anyone to give the money to, to complete the projects. And without those aid agencies you can't get development going, as they are the ones who do it.'

Brigadier Thomas's statement was a powerful message for the Foreign Office and was considered 'a good thing', for his brigade's move to Lashkar Gah demonstrably showed that here was a 180-man headquarters that planned ahead and held all-embracing meetings. To begin with, the local government and the British representatives in the Provincial Reconstruction Team pondered over 'this machine that has arrived'. Members of both 'camps' lunched and dined together while government representatives were invited to all the brigade's planning sessions – proving that they weren't trying to own the area but that they saw themselves as part of the collegiate whole.

Soon after arriving in Afghanistan, in September, Colonel Holmes had deployed his men into some of the key positions in

Helmand; yet there remained one last stronghold for which he was responsible: the town of Now Zad. To the north-west of Helmand Province, this once-thriving market town nestles on the fertile valley floor between ridge-backed, stony mountains and is fringed to the east by dense cultivation and even thicker stands of trees. Outside the town, numerous farming compounds serviced by kilometres of irrigation and sewage ditches break up the otherwise barren landscape.

On his arrival, Matt Holmes had inherited an infantry company in Now Zad. In the middle of the town the Gurkhas had, earlier and gallantly, defended the District Centre, an old police station; there had been heavy fighting, including hand grenades being exchanged over the compound walls. The Gurkhas had, apparently, even drawn their kukris when the occasional Taliban got in. Overlooking this building and covering it by fire was ANP Hill, a high mound close by the south-west of the town. It was imperative that that position be held – no matter what, and not just because of its commanding view over the town: behind ANP Hill, to the west and out of sight of much of Now Zad, lay the helicopter landing site, the vital link with the outside world.

Matt Homes was not happy with the situation he inherited. 'Alpha Company of the Royal Regiment of Fusiliers, under Major Jonathan Swift, had been there nearly three months – in the end they completed 110 or 111 days – in about as tenuous a position as Rourke's Drift and they, understandably, felt somewhat neglected. Living conditions at the District Centre, with little or no overhead cover, was so appalling that they had to be improved as our first priority.'

Once again Colonel Holmes had a simple answer to some of the problems Alpha Company faced: offensive spirit. 'They had been given that District Centre to defend, they were not allowed to leave the compound, and that was all they were tasked to do. I told them to go and disrupt the Taliban and deny him freedom to manoeuvre in the area. Alpha Company were thrilled and swiftly conducted a couple of raids to destroy firing points near the District Centre that had been giving them considerable trouble. It did wonders for their morale.'

Colonel Holmes and his then regimental sergeant major, Marc Wicks, together with Major Sutherland – whose Kilo Company, after spending a few weeks in Gereshk and then Bastion, was to relieve Alpha Company at the end of October – had flown in for a recce and spent three days with Swift's gallant men. 'The fusiliers were doing a stoical job there. A close-knit company.'

With his commandos now settled into the places where he believed they would be most effective, and with the men adjusting to the heat, dust and general conditions in Afghanistan, Matt Holmes was able to offer an interim summing-up: 'We were told to hit the ground running, which we've done, having had successful engagements with the Taliban within days of arriving. There is plenty to do in support of the Afghan authorities and people, but it will take time. We are robust, highly capable and have unrivalled firepower at our disposal. We're all looking forward to the challenges ahead and I don't expect there to be a dull moment.'

2
OCTOBER:
COURAGEOUS INACTIVITY

'What made it difficult was not the incoming fire but moving through the saddle of the mountains along a track that was only half a metre wide: all that separated us from the Russian mines. At night with no light the route was hard to see, especially when moving tactically, under fire and carrying heavy loads.'

Of all the places that Royal Marines had visited over the years, the one they least expected to resemble Northern Ireland was Helmand, but it wasn't the geography of the place that seemed to resonate in their minds; it was the similar experience of bringing peace to a people who, even when not overtly siding with the Taliban, reacted to the British as if they were hostile invaders. It was this, and the interweaving of personal and political matters in the day-to-day lives and relationships of the people who lived there, that most struck them in its similarities to the now-peaceful province. Even the bases some of them operated from reminded them of Northern Ireland: 'FOB Price was a patrol base just like Crossmaglen.'

On 8 October 2008 3 Commando Brigade's brigadier, Jerry Thomas, formally took over from 16 Air Assault Brigade's brigadier, Ed Butler. While Brigadier Thomas wished to build on the successes of his predecessor and maintain a form of campaign continuity, he also needed to move beyond the 'break-in battle' philosophy that had so dominated earlier thinking and that had effectively, and necessarily, been extant throughout the Paras' time in Afghanistan. Across the province, his units were settling into their tasks, taking over 'the footprints', understanding their local areas and considering how to act out his instruction to be 'dynamically unpredictable' through manoeuvre. Above all this, however, was a legacy that had to be addressed by the brigadier; he wanted to 'un-fix' the brigade from static locations. Nowhere was this more apparent than at Musa Qal'eh.

Musa Qal'eh's problems had begun long before the transition of authority and, for some time, had severely tested the resolve of Brigadier Butler and his 16 AAB. Musa Qal'eh was, too, a nightmare for 3 Para's Easy Company, the resident troops, and if the intelligence was correct, about to get worse. It was an important place not just for its position – towards the north of Helmand, it was close to the centre of Afghanistan, and closer to the dam at Kajaki – but also because the Taliban used it as a staging post and the place where, after the winter, they would regroup for the 'spring offensive'. Denying them the use of the town would not only prevent them launching attacks on the dam and other reconstruction projects, it was also a vital part of the effort to clear the Taliban out of northern Helmand.

This explains why Musa Qal'eh was important to the insurgents, and why the battles fought there had created so many difficulties for the local inhabitants. For the Paras of Easy Company it was no better; there was only one way in and one way out of Musa Qal'eh and that was by air. The Taliban could easily target these aircraft; already there were a growing number of small-arms strikes and close calls from the ubiquitous RPG. The position was becoming untenable and a fast-rising casualty list drove this point home. By 8 September stocks of ammunition were so low that, for instance, enough mortar bombs remained for only one more major engagement. There were unconfirmed reports that four or five hundred men were now camped two and a half miles from the town.

Given that there were no resources to hold the town, and that it no longer fitted with the strategic decisions of the brigadier, there were questions about continuing to hold Musa Qal'eh at all. What was the point? The point was that to leave would hand the Taliban both a moral and a physical victory yet, balanced against that, was the real chance that a disaster was waiting to happen: a disaster such as the destruction of a fully laden Chinook – a 'victory' that would be trumpeted internationally by the enemy.

In the end the decision to call a cease-fire and end what was, in reality, a stalemate – with both attackers and defenders worn down by the attrition rate – was made on 12 September, not by the Taliban nor by the British but by the town's elders. One of the terms of the ceasefire required both protagonists to withdraw, providing no hostilities broke out for a month. From Easy Company's perspective, this was not what they wanted; to stay put was the correct solution, for it meant that those of their number who had died had not done so in vain.

'It was an Afghan solution and, if that is what the elders wanted, that is what we would do,' said Colonel Matt Holmes. 'On the basis – their argument – that if you remove the target the fighting will stop.' Holmes – whose task it would be to ensure that Musa Qal'eh was relinquished in good military order – had his own views. 'That is what the elders thought and we had enough confidence in them to say "yes" to the operation: they were quite strong with the Taliban then.'

Yet it was also clear that, with a vacuum in Musa Qal'eh and after a suitable absence and at a time of their own choosing, the Taliban would return; then the British would have to fight their way in again. This is exactly what happened just over a year later when, as part of an Afghan/ISAF force, 40 Commando Royal

Marines forced their way back to the centre of the town. From the military point of view any withdrawal from the beleaguered Platoon House was bound to be a short-term solution.

42 Commando Battle Group's first offensive task – *Operation Omer* – was planned for 13 October and was to provide force protection and an 'on-call reaction force' if anything went wrong during the agreed extraction of 3 Para's Easy Company from Musa Qal'eh.

'Easy Company,' Holmes says, 'was predominantly Royal Irish – I think there were only about six or seven Para soldiers – commanded by a Para officer, Major Adam Jowett, a good guy whom I spoke to a lot. Musa Qal'eh had seen some of the fiercest fighting with a number of fatalities, and even the Pathfinders – the reconnaissance unit of the Paras – had been fixed there for quite a while. Anyway, the elders approached Adam and said, "We will guarantee you a safe passage – as much of a guarantee as you will get and not a written guarantee – if you leave." The decision had been made.'

Next, Holmes brought Kilo Company, with Neil Sutherland and his HQ, to a point in the desert near an old Soviet defensive position, although they kept a reasonable distance away because of the possibility of mines still lurking beneath the rocks and sand. Together with a couple of Chinooks and a battery of 105 guns from 29 Commando Regiment, Royal Artillery, Holmes believed he had 'enough combat power and reserves to provide an assault force into Musa Qal'eh if we had to go in to extract the company as they drove out'. He knew there were Taliban in the area, so it was a high-risk operation. 'But I spoke to Adam and we came to the view, based on his knowledge of the area, that it would probably be OK, that it was a "goer" and on that basis we agreed the operation.'

Orders were given that evening and everything prepared for the extraction. Then, in the middle of the night, the town elders came to Major Adam Jowett and told him that they were going to change the extraction route for Easy Company. This was perceived as an unhelpful move, and one fraught with even more dangers than before, especially as part of the ceasefire agreement meant that Colonel Holmes was forbidden to put any combat power into Musa Qal'eh itself. The plan now was for Jowett's men to be driven in a convoy of 'local' lorries – Afghan 'jingly' lorries, so called because of their bright ornaments and colourful paint schemes – out of Musa Qal'eh, but no longer straight across the wadi into Kilo Company's protection, the shortest, quickest and most easily picketed and defended route – although a small amount of planned clearance on the wadi's western side by engineers would be needed. Now they were to be driven down Musa Qal'eh's high street into a complicated wadi system before turning for safety.

Holmes's real concern was for the safety of Jowett's men, but he was also worried that this might have been some form of publicity stunt, with British troops being humiliated in public – followed by pictures splashed across the world's media. Real danger also lurked in the two wadis, as both were mined and both had been used as killing areas where men had died only recently.

Well-founded though they were, the colonel's fears were not realised, as an elder climbed up into each truck before they set off. 'Easy Company went through about six kilometres of wadi, led by the elders. This gave me a degree of assurance despite the "jinglies" having hardly any protection – just a few sandbags on the floor and that was about it. It was my first battle group operation and very high risk with much of Whitehall watching as it unfolded. No shots were fired, which was significant because everything inside the District Centre had been tense and previously, of course, the scene of heavy fighting.'

With Easy Company safely delivered, Holmes – accompanied by Jowett – held a *shura* with the elders in the desert, during which he gave them water and footballs – which caused some amusement – expressed his appreciation for their part in the undertaking and wished them well in the running of their town.

One of the suspicions of the British had been that the Taliban often came to these meetings, to look their enemy in the face, as it were. It was difficult to tell; all men wore the ubiquitous *dish-dash* or long flowing cloak and sported turbans or *shamaghs* of varying colours.

Neil Sutherland – with his commanding officer and everyone else – was relieved by the peaceful outcome. 'We were glad it never actually came to a punch-up as all the routes into Musa Qal'eh were heavily mined. The Paras ended up being taken way up to the north: we never knew why and it got us all a bit edgy, although we guessed it was because the elders knew where the mines and IEDs were not. Had we had to go in, the only way we could have done so would have been by putting our EOD – Explosive Ordnance Disposal – vehicles at the front and just putting our foot down and driving straight into the town. The Paras' camp was in the middle of Musa Qal'eh so it would have been very, very messy. Of course what we couldn't plan on was where we might have to go to rescue them, so we were constantly looking at the route as it changed. We assumed the Taliban were also watching, but it is difficult to tell an unarmed Taliban from an unarmed civilian.'

Easy Company flew home, grateful at last to see the back of Musa Qal'eh.

Kilo Company returned to Camp Bastion, but not for long. The continuing work of replacing the men in their fixed positions meant that Alpha Company, the fusiliers, in Now Zad, were to be relieved by Neil Sutherland's men on 30 October. At the same time Captain Luke Kenny's 5 Troop was despatched to relieve 10 Troop at Kajaki Dam. The dam was one of the most strategically important locations of any held by the commando brigade: at the northeast corner of Helmand Province, the dam, when working to full capacity, is the major provider of hydroelectric power to the province and irrigation for 650,000 acres of otherwise arid land.

Built in 1953, it is 320 feet high and 887 feet long, with a storage capacity of nearly one cubic mile.

Within 5 Troop, Corporal Al Hewett's 1 Section's task was to help man the two OPs, Athens and Normandy, on the ridge running west–east to the south of the Helmand River, an area where a number of British and coalition forces had been killed defending static positions. The other two sections, 2 with Corporal Tom 'Webby' Webster and 3 with Corporal Den Dennis, had arrived a few days before.

Almost as soon as Hewett had moved in with his section, they were in action. The day after arriving, his troop was tasked with carrying out a reconnaissance of the westernmost peak, Sparrowhawk, and conducting a sweep for any signs of enemy activity. The peak rises to just over 1,100 metres at its highest point, and overlooks, to the west, a small series of compounds and – thanks to their proximity to the Helmand River – lush fields. 2 Section under Corporal Webster was point section and almost immediately on approaching the summit were shot at and mortared by the Taliban, who clearly regarded this as a strategic location not to be given up. For over four hours the section was pinned down by small arms, RPGs and mortars. Al Hewett considered it 'a deliberate ambush during which "Webby" did a great job of recovering his men under fire out of the initial area with covering support from my section, the Fire Support Group and the mortar fire controllers. To do this we had to neutralise numerous enemy positions, which involved all the lads giving timely and accurate fire control orders. These assisted Captain Kenny as he brought in the strike jets and attack helicopters.'

Within two days of that operation, 1 Section was ordered to conduct the same task again, but now with the knowledge of what lay ahead. The night before, Hewett's men were tense but rallied by their second-in-command, Lance Corporal Tom Birch, who had acted so calmly under fire at Gereshk the previous month. The section commander was under no illusion what his team would be facing, but he also knew that after the Gereshk experience, and the months of training they had undergone in the UK, they would not let themselves down.

The patrol set off in the early-morning light and almost immediately, as expected, came under sporadic but accurate fire, forcing Hewett to decide that it was 'too unhealthy' to hang around for long. With Tom Birch commanding a three-man fire team providing cover, Hewett and the rest of the section scouted forwards along the rocky ridge, careful not to step into the legacy minefield that had given 3 Para so much concern – it was laid right up to the edge of the rough path, and was totally unmarked.

The aim of the reconnaissance was to establish the Taliban lying-up position so, having achieved this, 1 Section, still under fire, withdrew back to OP Athens, where they were able to pass on useful information about the enemy's dispositions.

Despite these reconnaissance patrols Hewett was also conscious of other aspects of leadership. 'With the section static in one area for weeks I was aware of the men losing basic counter-enemy ambush techniques while patrolling.'

To address this concern a training package was drawn up which involved moving off the mountain to the Forward Operating Base at the bottom of the peak, where a 'patrol-shoot' exercise was planned in a relatively safe area that had not been attacked for some time. This involved the section, under Tom Birch's command, practising anti-ambush drills along a 200-metre track. Once they were on the track, however, the Taliban – who had, in practice, set up an ambush – attacked. Instantly the drills now became the real thing and the enemy was successfully rebuffed, but the Taliban must have been surprised at the speed with which they were countered, not realizing that the team were specifically training at that moment for this very eventuality. As it was, no casualties were taken and the section safely withdrew from the 'exercise killing area'.

Unfortunately this was followed by an accident that cut short the deployment on *Operation Herrick 5* of the gallant Lance Corporal Tom Birch. Corporal Al Hewett was on the top of the peak when he received a radio message, informing him 'of an accident involving Tom Birch and Matt Fenton somewhere between the summit and the Forward Operating Base. So I immediately organised a team to go and help. Luckily, the night before I had packed an emergency crash bag of medical kit, strobes and air marking equipment.'

North-east from the observation post on Athens a very narrow, rough and stony track winds steeply down to the river and Forward Operating Base Zeebrugge: either side are uncleared, legacy minefields. Towards the bottom the Land Rover, driven by Tom Birch, had left the track and plunged down a fifty- to seventy-foot cliff.

'The crash site was not pretty and the smell of diesel was everywhere. A number of lads were already trying to detect vital signs, with Tom needing the most urgent attention, although Matt needed it too. More men were appearing, including the medic. We placed a neck brace on Tom, started to clear his airway and checked his breathing and circulation.'

Tom Birch was in a bad way, drifting in and out of consciousness, bleeding profusely from the head, clearly with internal chest injuries and still in the vehicle, pinned down by the steering wheel. The team of helpers were 'finding it difficult to keep him alive'. Two army personnel, both with medical experience, clambered up the hill to report that a helicopter was on its way from Camp Bastion. Marine Dave White, another 1 Section lad – and one with enormous strength – prised the steering wheel off Birch, who was then inched free and placed on a makeshift wooden stretcher. Someone stabilised his spine while Hewett remained at his head, pumping blood out of his mouth and keeping pressure on the injury. All the while two navy medics were controlling the accident site, collating the vital signs of both casualties and passing the information to the helicopter crew and the on-board surgical team.

Al Hewett insisted on accompanying his two men on the thirty-minute journey back to Camp Bastion, where he stayed the night, 'occasionally popping in to the field medical centre to check for news on the two lads', when he managed to extract mixed and often grim prognoses. Birch was to be flown to Oman that night, where a specialist team from Iraq would be waiting. Luckily Tom Birch's life was saved, in part thanks to the swift actions of his comrades in the field, and although he is yet to make a full recovery at the time of writing, he's well on his way to returning to his comrades in the commando; Al Hewett included him in his group at Buckingham Palace when he was decorated with the MC.

'Covered in blood, I returned to the transit accommodation, where I met a few of the company lads on their way to the UK for R-and-R and although it was good to see friendly faces I broke down.'

The following morning Hewett made his way to the galley for breakfast; after some time in the field and looking forward to a proper meal, he sat down at a table. He had travelled straight from the crash site still in the clothes in which he had tended his wounded comrades, so as he sat in his heavily bloodstained uniform, stinking of diesel, sweat and the Afghan bush, he was approached by a chief petty officer, 'who was clearly distressed. I thought that he had come to give me bad news about Tom but instead told me off for not having clean clothes on. I smiled. He looked at the blood, got the picture and moved on.'

Before he returned to his section, Al Hewett visited Matt Fenton, to find him stable, unconscious, heavily bandaged but alive. He then caught a lift on a 'routine' Chinook flight to Kajaki, where he found that, in his absence, the fighting had intensified. 5 Troop, with more marines now on R-and-R, no battle casualty replacements and manning the two mountaintop positions – Athens and Normandy – was dangerously overstretched.

The Fire Support Group came under sustained attack night after night, proving their worth every time by returning fire though, as a result, they were starting to run short of ammunition. After Tom Birch's experience of driving off the cliff, understandably no one was enthusiastic about moving about on the edge of a minefield in a Land Rover loaded with ammunition in the middle of a firefight. But the marines, as ever, had a plan; and thanks to their extraordinarily arduous training, were fit enough to carry it out. Corporal Hewett took three men with him and ferried supplies from one OP to the other, loading ammunition belts over their shoulders and carrying boxes in each hand. The track they took threaded its way through the old Russian minefield, along the saddle, while the marines were only able to see where they were going thanks to their night-vision equipment – any lights would have alerted the Taliban to their position and they'd have made easy targets. Despite little food and tired bodies, the marines kept up a steady supply of ammunition, often carrying upwards of ninety pounds each. Fitness, stamina and determination played a vital part.

When the shooting died down, the view from the top of Normandy, particularly in the early-morning light, was magnificent: the vast reservoir, penned in by the dam that the Taliban were so keen to control, would reflect the clear blue sky above the bare hills that ran steeply down to the still waters. Nights were cold, but the sun warmed the men almost as soon as it rose, and that would be the time to replenish supplies more safely.

But after a few weeks of this it was clear that 5 Troop were going to need reinforcements if they wanted to clear the OPs to their west. In early December the whole of Mike Company would be 'surged' into the Kajaki area of operations with orders to push the Taliban back and, as a precursor to attacking compounds to the north of the river, to secure and hold the most vital of the mountain peaks,

Sparrowhawk East and West. Once these were taken, the Royal Marines would enjoy total dominance of the high ground around the Kajaki area by giving themselves a four-kilometre buffer zone from within which they would hit the Taliban.

Meanwhile and back in time, another of the traditions of the brigade had to be observed: just as an RSM handover had to take place during the operation, so too did the annual marking of the corps's formation, regardless of extenuating circumstances. At Camp Bastion, on 28 October 2006, the 342nd birthday of the Royal Marines was celebrated. 42 Commando had, as commandos tend to do on operations, brought with them both their Queen's Colour and their Regimental Colour, so these were paraded in front of those Royal Marines then in the camp, during a short service. There was even a rudimentary cake, in garish red, yellow and blue, with fifteen candles. This ceremony in the middle of a bitter conflict might have been considered unusual to an outsider, perhaps even eccentric during an operation such as *Operation Herrick 5*, but not to the Royal Marines, for whom it is a tradition staunchly upheld every year. It is from such observances that the spirit of the corps is enhanced, and it is this spirit which helped the men through some of the darker times in Helmand Province.

Morale might have been a problem for troops holding Now Zad earlier, but with the involvement of Kilo Company under Neil Sutherland, this was no longer the case. 'My company, less 5 Troop, flew into Now Zad, where we combined with one or two elements of the ISTAR – Information, Surveillance, Target Acquisition and Reconnaissance – group. Unfortunately, Jonathan Swift's fusiliers had never been allowed out until my colonel took charge of them.

'The fusiliers had been in an entirely defensive position, prevented from being proactive. They couldn't interface with the local population and so they were in quite a sorry state. That is no detriment to them: given the circumstances they did a very good

Typical Afghan landscape between Kandahar and Camp Bastion.

OPs Normandy and Sparrowhawk from Athens. Overwatch above Kajaki dam.

Relief in place.

Garmsir Bridge.

FOB Price – Gereshk.

42 Field Engineers on Operation Slate 1, filling Hesco Bastion containers.

Major Andy Lock briefing Zulu Company, 45 Commando, prior to Operation Glacier Five.

Major Ewen Murchison at a shura on MOG North. Mostly Taliban in the background.

Vikings arrive at Kandahar on Afghanistan civilian 'jingly' trucks.

Chess – Helmand-style.

Captain Will Mackenzie-Green batting at the Kajaki cricket match prior to operations.

Heads! FOB Dwyer Garmsir district.

Basic accommodation, Garmsir.

ANP Hill. Now Zad. Improvised bread oven.

Marine Chris Fraser-Perry (on the right) gives thumbs up before departure for Jugroom Fort.

Apache on way to Jugroom Fort with Captain Dave Rigg holding on tightly.

Chinooks landing at Sangin District Centre. Protected by a fire plan, own defence systems, Apache, Harrier and 105mm guns firing from FOB Robinson.

Now Zad main street.

2 Troop follow a blood trail after a fire fight near Gereskh.

Scimitar of C Squadron, Light Dragoons.

Desert Hawk being launched.

Pinzgauer.

Hill 836 immediately prior to 107 rocket attack. Gereshk. Operation Slate.

job. Even being fired at was more dangerous than it should have been for they had no overhead cover, the gate to the District Centre was a sheet of "wriggly tin" propped up by a Land Rover and they had no engineer resources.'

The most soul-destroying thing for the fusiliers in Now Zad, apart from the living conditions, was not being allowed out until Matt Holmes took over their command. They were thrilled, as they had been treated as second-class citizens by the Paras, whose last offensive operation, back at the beginning of June, had ended inconclusively. 'So the fusiliers left and the reins were now off. I reckoned I had a pretty good remit – bit of a blank canvas – as no pattern had been set and no promises had been made to the elders.'

Sutherland had much ground to make up, both metaphorically and practically, for six months previously Now Zad had been a thriving market town. Now it was empty: effectively a no man's land of about one square kilometre from which the District Centre was attacked on a regular basis from very close in, sometimes as close as fifty metres, via the narrow, almost secretive alleyways that surround the DC. The enemy had had complete freedom of movement.

The base needed much reconstruction; for example, as there was no running water the marines relied on bottled water for drinking. There was a well within the DC, which they used, rigging up solar-cell showers: black bags filled with well water. Without a purification system this water was used only for showers and the bottled water for washing up. Food was monotonous; they were on twenty-four ration packs to begin with – boil-in-the-bag stuff and hexamine cookers – until issued with ten-man ration packs – and a Royal Marines chef. 'We should really never need to go more than a couple of weeks on these rations without some supplements, but we never had any,' said Sutherland.

Accommodation was pretty basic too. Some men rigged up cots with mosquito nets against walls, but for many it was an underground room, a cell really, constructed of dry mud and only a mat to lie on. At least the roof kept them safe from mortars; but they met their fair share of problems down there, with the local creepy-crawlies putting in an appearance rather too often. Spiders were one thing – the size of a man's hand – but at least they were harmless. The scorpions, *Androctonus amoreuxi*, though, were a different prospect; known as 'fat-tailed scorpions' they are ranked among the deadliest on earth. Their venom is enough to kill a human and, being sand-coloured, they are capable of disappearing against the sand and rock.

Every creature, human and animal, sought shelter somewhere as the centre of the town had been devastated by the attacks launched by both sides. Most of Now Zad was made up of single-storey small compound houses, perhaps a tree in the back to provide a splash of green against the monotonous grey-brown mud walls everywhere. But the centre had been full of two-storey houses, with shops underneath, some of the houses with metal-worked, first-floor balconies; all now rubble, the bricks that had been covered by the mud strewn about in untidy heaps.

When Kilo Company arrived in Now Zad at the end of October they found that the local population had moved two kilometres to the north, where a new village and market were established. That meant Sutherland could initiate an aggressive patrol programme of concentric rings that, with each successive patrol, encompassed a wider and wider area. The marines began to understand the countryside in greater detail, including the likely firing-points from which the District Centre had been engaged on a regular basis.

A number of restrictions, though, were placed on the company commander. He was not to enter the wadi that runs north to south to the east of the town. Initially this was a restriction that the marines didn't mind as they became used to the other areas, but then pretty quickly it began to grate with them. The wooded area was close country and riddled with tunnel systems; the known fact was that the Taliban lived in the area, and from it they dominated all the villages and mined all the passes. It was clearly worth investigating.

Throughout their tenure in Now Zad – and as had been the case prior to Kilo's arrival – a Fire Support Troop was positioned on ANP Hill to provide over-watch of the District Centre. Behind this mound, to the west, Chinooks could use a makeshift helicopter landing site, mostly but not completely out of sight of, but not necessarily out of range from, Taliban positions to the east of the town.

To the town's east, beyond the irrigated fields, wadi and woods, lies a line of dragon-backed hills that divide the Now Zad valley from that of Musa Qal'eh. This major feature – named Crocodile Ridge by the British – runs roughly north–south and is passable via three narrow defiles that link the two areas. It was to be in one of these passes that Juliet Company was ambushed later in the deployment. Before that, though, Major Sutherland's eastern – right-hand – boundary was the out-of-bounds Now Zad wadi, a place of considerable interest to the company commander. 'As soon as we ever got close they would engage us in such strength that I never had enough combat power to do much about it. It was also the area where the Paras got a bloody nose in June 2006.'

History was, perhaps, casting a long shadow. 'Speaking to some of the locals that I eventually got to know, the Russians did not rid the area of Mujahadeen with a brigade and so there was no way I was going to evict them with a small company. But we started patrolling, started to dominate the ground with standing patrols at night. This had not happened before and the locals were a bit taken aback. We were still in helmets as we were getting contacts every day, but generally they were not engaging our foot patrols although they were engaging us in the DC and on ANP Hill.'

Neither Neil Sutherland nor his marines knew why the foot patrols were seldom targeted, for the alleyways, which ran right up to the walls of the compound, were little more than a metre wide in places – a man standing in the middle could reach out and touch each opposing wall with his fingertips – while the surrounding buildings were riddled with extensive tunnel systems. Command and control for the company commander in such a contained maze was a nightmare, and almost any kind of ambush there would have been a disaster. It seemed that the same thought had occurred to the Taliban, who seemed happy to stand off and rely on attacking the District Centre at long range, using their RPGs and their heavy machine guns.

As soon as he had made his initial security assessments, Sutherland felt able to strike up a relationship with the elders by making it known that he wanted to discuss events now that there were ceasefires in Sangin and Musa Qal'eh. 'The word clearly reached them as they contacted me and we had a chat. Once I had built up a good rapport, the subject of a ceasefire came up; I made it quite clear that I was not going to be fixed in our District Centre for, during a ceasefire, there was so much we could achieve. Projects such as regenerating the medical centre, right alongside the District Centre, and clearing some of the rubble off the streets. We had a team of 59 Independent Squadron, Royal Engineers, with us who were itching to get stuck in with that sort of thing. The closest we got was a statement that the locals wanted the Taliban to stay where they were and the British to stay where we were.' Which was as good a ceasefire agreement as he was likely to get, especially as he was cleared to conduct 'framework' patrols.

After his extensive experience in Northern Ireland, one thing was apparent to Sutherland in Now Zad. 'Patrolling among the local population should provide security but in Helmand all they were achieving were fights with the Taliban among the population – and then the locals get pissed off.'

At last, Neil Sutherland was able to report to Colonel Holmes that there was an agreement for a ceasefire; and it took place several days later. The elders had negotiated it with the Taliban and with coalition forces through the local governor. All were happy, especially the British as they could continue to patrol, but everyone knew that it would break fairly rapidly; in the end it lasted about two and a half to three weeks. During the ceasefire, the marines continued to receive 'incomers', but these were relatively short contacts and engagements. Sutherland's view was that 'it was vital we did not overreact. I had to explain to the young marines that just because their Rules of Engagement said they *could* open fire it doesn't necessarily mean that they *should* open fire. "If you open fire the ramifications could sometimes be far greater than those resulting from not opening fire," which is what the brigadier had meant by encouraging "courageous inactivity".'

Kilo Company could now expand its patrolling programme north to the 'new' bazaars and habitation, then across the open ground, rough desert scrub and individual compounds to the north-east in a huge loop around the area; all the while linking up with the locals they met as far out as the Taliban's wooded retreat in the east. One of the irrefutable arguments offered by the civilians they met along the way went like this: 'You know where the Taliban are, you know they are over there, so why do you not do something about it?'

In reply to this perennial truth, Sutherland, as with all other commanders across Helmand, had to be careful when answering these accusations: force levels simply did not allow for such sweeping action at company level. 'The truth of the matter was, we could only do so much with the men we had and we could never make a promise.'

Colonel Matt Holmes held the view that the base at Now Zad was 'not strategically important', and 'was the first one I would have given up had I been able to'. He goes on to admit, though, that there was value in the occupation of the District Centre by the Royal Marines, especially if one treated it less like a static base and more the launch pad for mobile operations. 'During Kilo Company's tenure, when we really started to take the fight to the enemy as well as build relations with locals, including the elders, there were occasional periods of cessation in hostilities, which we exploited by conducting civil military cooperation tasks to garner local support and demonstrate our good intent. The Taliban, recognizing a threat to their own aspirations and influence, resumed fighting to try to deny us this avenue of approach.'

Meanwhile at Sangin the colonel had felt able to reduce the garrison by 50 per cent and to task Royal Engineers of 59 Independent Commando Squadron, under Major Nigel Cribb, to help with improvements. His own Assault Engineers, commanded by Colour Sergeant Stu Patterson, 'were carrying out one hell of a lot of overdue work to improve the defences and living conditions, including sanitation'. During their time at Sangin, the Paras had had about 200 men stationed in the DC, but then they had been in frequent contact with the enemy with little time to devote to 'domestic' issues.

At Gereshk, and across the same areas that Kilo Company had been in contact before handing over to their successors, a more mobile form of warfare was developing. In keeping with the aims of both the brigade's commander and Lieutenant Colonel Matt Holmes, Juliet Company under Major Ewen Murchison, in FOB Price since 6 October, was tasked with developing a unit similar to those made famous during the Second World War in North Africa, a long-range desert patrol group.

As with Kilo Company in Now Zad, patrols were despatched to demonstrate among the local Afghanis the enduring commitment to actively improving the fragile security situation. The company regularly supported local ANP patrols and conducted Quick Reaction Force operations to deter and defeat attacks by the Taliban; it also protected the town centre from being overrun, as part of its policy of providing reassurance for the civilians.

Once security in Gereshk had improved sufficiently, Juliet Company were in a position to carry out Combat Recce Patrols, which served a number of purposes. Mainly these were to ensure that the Taliban didn't have free use of the area, but they also added to their intelligence of the surrounding vicinity. As part of the process, the patrols would enter compounds and conduct searches for weapons or any signs of Taliban activity.

Outside Gereshk, Juliet's main task was to conduct desert patrols or Mobile Operations Group (MOG) patrols across Helmand Province, with particular focus on the Sangin Valley. Here, Murchison's orders were to 'find, disrupt and defeat' the Taliban, engage with the locals and conduct civil-military – CIMIC – activity wherever possible.

For these MOG 'excursions' the company group regularly numbered about 250 men and forty vehicles, which included a highly manoeuvrable element of fourteen Viking and four WMIK. They also had 81mm mortars, two 105mm light guns, Indirect Fire Weapon Locating Radar, UAV – the unmanned reconnaissance plane, Desert Hawk, that was launched by a large rubber band – plus the key logistic teams to keep the show on the road, fed, watered, fuelled and 'ammoed'.

The Fire Support Group of four WMIK and one troop HQ Viking were the 'eyes and ears' of the company group, while the arrival of the grenade machine gun midway through the tour significantly enhanced the company's firepower. Each of the two Close Combat Troops required three Viking, leaving six Viking for company HQ with Fire Support Team embarked. The forward repair vehicle and the medical vehicle were paired with the ammunition resupply vehicle: the mortars were paired with the second-in-command's Viking, responsible for identifying and securing a helicopter landing site in the vicinity of the mortar line in case the incident response team was required. Murchison preferred to hold the casualty-evacuation and ammunition-resupply Vikings centrally.

The company's operating area was split into three distinct zones, which could be separated as much by geography as by threat level. Furthest from the action and therefore the best place for any OPs that might need to be established was the undulating terrain at the foot of the mountains; with little habitation, it was ideally suited for laying-up positions, but these places were usually at the limit of artillery support. Next, the open desert, with its scattered settlements and numerous tracks, was somewhere that could be relatively easily monitored, thanks to the ISTAR aids available to the company, from the UAVs through Nimrod aircraft to satellites; but this very asset was also its biggest problem, as the company was just as visible to anyone on the ground and therefore prone to 'dicking', a term used in Northern Ireland, meaning that progress was watched and reported by passing civilians probably more sympathetic to the aims of the Taliban than to those of the Coalition Forces. Basic but effective methods of 'dicking' were evident through the use of mirrors and smoke signals. Once contact had been made, a key indicator of impending action would be the rapid movement of women and children away from an area soon after the arrival of Juliet Company, quickly followed by the appearance of white Toyota Hilux vehicles – four-wheel-drive pick-ups with five men in the front and, usually, a mortar base-plate and heavy weapons in the open backs – and motorcycles and scooters carrying men and weapons. When the Taliban did launch an attack, it was possible to gauge the sophistication of their operations by the co-ordination of their defences, the outer cordons and multiple firing-points that made the most of the micro-terrain. The individual Taliban soldier always displayed a willingness to counter-attack and maintain contact – until he were hit by air power.

Finally there was the 'green zone', a two- to three-kilometre-wide stretch either side of the Helmand River, littered with irrigation ditches, impassable canal systems, ploughed fields, hedgerows and walled compounds; this was often accurately described as being similar to the Normandy *bocage* that had so frustrated the D-Day landings in 1944.

Within this green zone there was a limited number of tracks which could be used as transit routes for vehicles, but driving down them brought the risk of 'channelling' – being directed against your will into an area which the enemy could use to launch an ambush – and were therefore designated as 'severely restricted' for vehicles and to be used only *in extremis*. Entering the green zone carried the inherent risk of contact as the Taliban enjoyed complete freedom of movement. It was soon discovered that there was definitely a line drawn on the outskirts of these areas that, once crossed by Coalition Forces, would trigger a contact from small arms, RPGs and mortars.

During their patrols, Juliet's MOGs covered a vast amount of ground, enabling them to gain a good appreciation of the terrain, urban centres and patterns of life. Still, they could never tell whether the compound or settlement they approached would be a friendly or a hostile one.

'It was a bizarre reality, but as you approached previously unvisited settlements, there was an equal chance of being greeted by the elders and invited in for tea, or being engaged from multiple firing points. This led the lads to coin the phrase "advance to ambush". The result was that we were routinely on the back foot and subsequent attempts to draw out the Taliban into killing areas of our choice were often fruitless.'

The phrase 'advance to ambush' became a favourite over the next few months, until it seemed to be less of a private joke and more of an official instruction.

There was always a temptation to take the war to the enemy more and more, but simple numbers usually ruled against this. 'Through engagements along the green zone's fringes we would have undoubtedly had an attrition effect on the Taliban,' said Ewen Murchison, 'yet it was equally clear that to have an enduring effect, discrete operations would need to be conducted *inside* the green zone. We did conduct limited operations within the zone, but only when sufficient resources were available. With the order of battle forced upon me by manpower constraints there was always a very real risk of being enveloped if we pushed too far.'

Juliet Company's standard approach was to establish a forward base for the Combat Support and Combat Service Support Groups which had to be within artillery and, ideally, UAV range of the target area. The Mobile Operations Groups would then use the umbrella cover this gave them to conduct their tasks, although the company commander would try to avoid co-location as much as possible, for the 'siting characteristics' of the Mobile Operations Group and support groups were mutually exclusive. The company's Fire Support Group would be employed to provide surveillance, to acquire targets and general reconnaissance.

Careful planning was needed to ensure that the significant logistic tail did not slow the MOG up; even when the moment came to begin battle procedure, plenty of time was essential, for the equipment – and therefore the personnel – required for a particular job routinely changed. Specialists such as the Explosive Ordnance Device Teams were as important to deliberate manoeuvre as everyone else, but for this capability to be effective it had to be deployed in vehicles with complementary mobility: the alternative was to position them with the support group, but if that were done the EOD teams would be too far back when needed.

One aspect of the procedure that all enjoyed was the regular monitoring of Taliban 'chatter', at least at a low level. The Taliban's radio communications, a simple 'push-to-talk' high-frequency system, were regularly intercepted and, thanks to the ease with which this was done, battle plans could be changed in the field at short notice. Obviously the Taliban were aware that their networks were not secure, but there was nothing they could do about this, other than – occasionally – using them as disinformation tools for the NATO forces. Intercepts ranged from those that purely raised morale, such as, 'We put down many nails – mines or IEDs – for them but they didn't stop; these are very hard people,' to those that assisted with the adjustment of indirect fire – 'Two mortar rounds just landed twenty metres from us' – to which the British reaction was always, 'Repeat the fire mission.'

Sergeant Nige Quarman of 2 Troop, Juliet Company, was involved in his company's first major contact when conducting a lengthy patrol in the eastern green belt of Gereshk on Wednesday 30 October. Due to under-strength manning, the two close-combat troops were operating with five- and six-man sections, the knock-on effect being that some marines were carrying a combat load, including ECM (electronic counter measures) equipment, of about eighty-four pounds; in addition to the thirty pounds of body armour and helmet.

Towards the end of their patrol, 1 Troop was in the lead, followed by Company TAC HQ, then 2 Troop, in a long, staggered line that did not present an easy target for the Taliban but that was close enough together to offer support if required. As a light mist rose from the edge of the Helmand River, the lead elements of the company reached the summit of Hill 852, an over-watch position to the east of the river that covered the dam and ANP guard posts. Here the marines settled down beside the CSM's quad bike – used for ammunition resupply forwards and casualty evacuation backwards – for an end-of-patrol 'banyan', an unlikely name for a moment to grab some rations and a drink. Then, just as the rear troop were about to start their final climb, a number of 82mm mortars landed near them.

As soon as the Taliban opened fire, everyone dropped to the ground, with those around the quad bike – 'big Bootnecks' – all trying to crawl under it. From their shallow shelters, members of the patrol quickly identified the Taliban's firing positions, further up the hill ahead of them. Swiftly the men in the WMIKs began firing back, laying down enough small-arms rounds to allow 1 Troop to push round to the side of the feature, from where they could observe the enemy manoeuvring to engage them. 2 Troop, the rear troop with the attached Assault Engineers and Sergeant Quarman among them, realised there was only one way to go, and that was up. Despite having completed a lengthy and exhausting patrol, the men stood up and, sprinting through falling mortar bombs, made it up the 300-metre slope to the top, shouting and firing as they went.

Amazingly, they reached the peak unscathed with Quarman among the leaders, alongside Marine 'Pat' Jennings, who was carrying a general-purpose machine gun. With lungs burning, they flung themselves down to start firing back at the scattering Taliban, but as they had no chance of bringing up Apaches in time, it was considered safer to break contact and return to FOB Price.

A salutary lesson had, perhaps, been learned early enough for it not to be repeated: the Taliban could strike at any time and particularly if a lessening of vigilance was detected. No more picnics around the quad bike, then.

3
NOVEMBER: THE STRATEGIC CORPORAL

'As we started to move forward we found what we thought was a dead Taliban who had been hit by an Apache 20mm cannon. But as he had mortar rounds underneath we weren't too sure if he had booby-trapped himself or if his "friends" had done so. As we were about to confirm that he was no longer a threat he moved a hand – so was killed in self-defence.'

Colonel Matt Holmes regarded the development of the town of Gereshk and the surrounding area as vital to his overall plan. The dam, to the north-east of the town, drew the Taliban in; determined to control the infrastructure that improved the lives of those nearby, they attacked the security forces guarding it. Through the use of thermal imaging equipment, it was possible for the Royal Marines to observe the enemy using the waterways – in particular the canal – to move towards their chosen targets. And not in a haphazard way, either, but with the precision of a well-drilled force: 'The Taliban were hand-railing down, walking alongside the waterways in military formation, staggered file and that sort of thing.'

To deal with the insurgents, the colonel's plan was to establish a permanent presence in the area of the dam in order to prevent the Taliban using the canal as a route into Gereshk itself, with the ANP manning the checkpoints that the British would establish.

Operation Slate 1 was devised to create these new permanent vehicle checkpoints, or PVCPs, so that there was a buffer zone between the town and the country beyond.

Under Holmes's command was Gereshk's resident company, Juliet; 45 Commando's Whiskey Company; Mike Company, the 'Ops One Company', held at thirty minutes' notice at Camp Bastion; and, under the protection of this compact task force, 42 Squadron, Royal Engineers, who would drive their plant and operate their diggers in full view of any enemy.

At the beginning of the day, Corporal Mick Slunker from Whiskey Company, in peacetime a drill instructor but now a section commander with 2 Troop, was surprised by his orders: 'On the way out we had been told that the enemy were waiting for us and we were to "advance to ambush". I've searched through all my notes since but can't find that order anywhere. Weird, that.'

In the dark of the early morning, on Sunday 5 November, at 0330, Whiskey Company loaded up their Pinzgauers and Vikings and set off towards Gereshk through the desert. It was cold and the noise of the trucks carried far through the still air, announcing their progress. They travelled in sixteen vehicles; spread far enough apart so that there was no chance of a mine affecting more than one vehicle, but close enough together so that one could support another if necessary. The threat of IEDs was present, and the Pinzgauers carried no ballistic matting to protect the marines inside against an explosion. The Vikings had yet to make an impression on the marines but that was soon to change.

The small convoy approached an ANP checkpoint, but as they drew closer, one of the Pinzgauers broke down. Sergeant Darren Stubbings, who'd been crammed into the back of the vehicle with six other men, dismounted and led his section to a collection of mud buildings close by, built into the side of the desert. 'I decided to investigate one and shone my torch to see what was in there, half expecting to see weapons, but there were just tyres and farming equipment.' Farming equipment seemed an unlikely item in an area covered in small sharp rocks, 'a pain in the arse for our knees', as well as a needle-like spiky grass.

The vehicle repaired, Whiskey Company continued on its journey down the Musa Qal'eh road into the middle of Gereshk. The walls of the compounds surrounding the dirt track started closing in on them where the track narrowed, and as the Vikings rumbled by, the crews and passengers were watched by people emerging from their homes, curious in the early light to see what the commotion was about. Children waved while others ran alongside; the older inhabitants looking the convoy up and down, trying to assess what the British were doing. The way ahead became more crowded, forcing the drivers to slow down in the constricted streets; the marines on top of the Vikings, and those exposed in the back of the Pinzgauers, became a little edgy; trigger fingers slid for the safety catch. Sergeant Stubbings told them to 'chill', but it was easier said than done.

'When we reached the bridge over the river, we turned left and headed towards the hydro dam, which is where the people started to disappear and where the lads felt much safer. Once at the dam the ANP were all around the place, so the marines went straight to their pre-ordered positions – well, kind of, as someone in HQ stopped us. I had to go up and tell them to crack on, which wound me up a bit. Once in our blocking positions a team of Royal Engineers from 59 Squadron erected barbed wire to secure our location and block any access route.'

The irrigation canal leading south-west from the sluice gate, through the dam and on past the town's southern limits, has a track on the northern bank's levee – a sort of towpath – which is raised some feet above the water's surface, wide enough to be traversed by a vehicle. It was along the edge of this track, towards the dam, that Corporal Mick Slunker led his men. Advancing carefully on the village nearest the dam, out of the silent morning air they heard a distant *crump* that the Corporal knew was a mortar firing in the distance. A moment later the tell-tale whistle of a bomb falling forced them to take cover before a deafening explosion showered dirt and rocks all over them. The round had landed about fifteen metres away from the section: 'When we came to, I checked everyone but because it had landed on the track, we were below the level of impact and were unscathed. I'll never forget Blaikie, the mad hairy Jock in my fire team, saying, "Mick, we need to get the fuck out of here!" The fact that you could hear the base-plate firing in the distance was eerie, then it was a case of counting the flight time and rolling the dice.'

The section had split into two teams, with one under Lance Corporal Mike Collins working its way through the compounds towards the dam. Collins clambered on to the roofs of the dwellings – 'You couldn't call them houses, they seemed to grow out of the ground they stood in' – to try and identify the enemy fire positions. Unfortunately for him, he was spotted on the roof and a mortar round came hurtling out of the sky, crashing through the roof – only it was a 'blind' and didn't go off. Someone was looking after him that day.

The mortar bombs were landing about them now; over the next twelve hours they didn't stop. The men learned to judge as best they could where the bomb would fall, counting down the seconds from the initial, distant thump in order to gauge how far away the Taliban were.

The aim of the Taliban was better than the condition of their munitions. Rob Lee, the Minimi gunner – a belt-fed light machine gun – had taken up a position on top of another roof for a better arc of fire down the track when a mortar landed in the open ground beyond the building. When the dust cleared Lee was nowhere to be seen, but just as his commander was thinking the worst he suddenly appeared, having been blown off his perch, running around the corner shouting, 'Mick, fuck staying on that roof, mate,' before jumping down into the wadi. As soon as his feet hit the deck the Taliban, who had moved closer to the patrol under their covering mortar fire, opened up with their AK-47s from about 300 metres. Mick Collins quickly organised his men into retaliation against this encroachment on their right side, and ordered them to return fire. 'It was weird giving a live-fire control order. I'd done it before a million times on Woodbury Common, Dartmoor or wherever, but this was the real thing. What was brilliant was seeing my whole fire team react to it without hesitation – humbling, and a credit to them.' The marines were doing what they'd been trained to do, and, after coming under fire from the Taliban, were only too happy to give him a taste of his own medicine. So keen, in fact, that Collins's GPMG gunner, 'Lynchy', stood up to fire from the shoulder.

After two or three minutes one of the company snipers, Jamie Sanderson, turned up to offer his assistance. Rob Lee was on the levee with his Minimi, a perfect weapon for these conditions with its effective range of 300–1,000 metres. He was kneeling down, right leg forward, bracing the gun against his right shoulder and squinting down the sights, squeezing off shots. Now Sanderson plonked himself next to Lee, rested his sniper rifle on his arm with the barrel of the .338 alongside Lee's head; isolating his target – a Taliban fighter clutching an AK-47 – he let off a round. Lee flew backwards from the levee, flipping in mid-air to land on his stomach, and lay still. The men around him thought he was dead, shot in the head. 'Medic!' one shouted, before Lee turned over, spat dirt from his mouth, sat up, and said, 'Fuck me, twice now!' Sanderson's job, though, was over; the Taliban retreated from their positions and the men were able to resume their slow progress forward.

Moving through the compounds was a good way to close on the target; however, in one compound they came up against an unexpected obstacle: a dog, frothing at the mouth and acting oddly. Collins – mindful of the efforts they were making to keep the locals 'onside' – radioed to ask if he could shoot it. 'Your call,' he was told. He looked at the dog and realised it was rabid, so there was only one thing to do: he shot it, as cleanly as he could.

As well as the earlier missiles, rockets now started landing among them, killing one of the Afghan National Army soldiers and seriously injuring an Afghan National Policeman with an RPG. Using the walls of the compounds as cover, the marines were able to return mortar fire, with 'Tommo', the company clerk, firing the 51mm mortar to good effect: twelve rounds with high-explosive warheads were returned, killing at least one. These HE rounds were popular with the marines, who were disappointed they were soon to be phased out of use. After that it went quiet for a while. The CSM issued instructions for the wounded and dead Afghans to be gathered up and placed in the trailer of his quad bike so that they could be transported back, via helicopter, to Camp Bastion for treatment as appropriate.

The men were able to establish the predetermined position for one of the PVCPs. A cordon of barbed wire was thrown around it and the Royal Engineers started work. After some time, the Taliban returned, with more mortars and artillery rounds into and around the commandos' area. A 107mm rocket punched its way through the roof where a marine was observing enemy fire positions but luckily it too failed to detonate. The JTACs – Joint Tactical Air Controllers – alongside the marines called up 'fast air' and talked them in to bomb the Taliban positions. The marines stood and waited for the bombing to take place, but were caught out by the speed of the planes – and the accuracy of the JTACs in guiding them in – so that the bombs had struck their targets and sent columns of dirt high into the air before the sound of the aircraft had even reached them.

While this was happening around them, the Royal Engineers were quietly cracking on with the building work, disarmingly oblivious to the dangers close by. One marine observed, 'We had stacks of admiration for them, driving around in their "diggers", getting on with the job. The day wore on; more rounds landed and after a while we lost count.'

The main barrier used in a PVCP is sand, which the Engineers collect in their 'diggers' and lift into large, boxy-looking, open-ended rectangular sandbags called Hesco Bastion containers, each about man-height, which stretch out in concertina fashion within a wire frame: this is why they're sometimes called 'concertainers'. These chains of empty boxes are delivered flat, then quickly pegged out to form the wall; they are the best way of achieving protection. Once the heavy-duty plastic liners have been filled with sand, they make a good defence against most forms of direct attack, and in a dull khaki they blend in well against the dirt around them. Observation-cum-machine-gun posts were established by the engineers driving large wooden posts into the ground, making a rough frame on top of which a corrugated-iron roof provided overhead shelter. If there was room behind the sandbag wall, pallets would be shoved into place to form rudimentary seats.

The pattern of the day established, it continued in much the same way; sporadic attempts by the Taliban to disrupt the work of the marines and sappers were forcefully repelled. A 'spotter' was seen in the trees, signalling with a mirror to direct the mortar fire; he was shot. In all, twenty-three Taliban were killed that day. The evening stretched out and still the occasional *crump* of a mortar or the chatter of small-arms fire could be heard, but no further casualties were taken. 'As night fell, I thought how lucky we had been and how well the lads had done,' said Sergeant Stubbings. 'The Taliban were firing 82mm mortars and 107mm rockets, so we were fortunate to have taken no casualties. As the hours wore on, we regained the initiative and were able to hammer the enemy, despite being in our reasonably static locations. At the end the Royal Engineers of 42 Squadron were the real stars of the show. They behaved supremely well under fire and just carried on.'

Throughout the night, engagements with the Taliban continued across both Juliet and Mike Companies, causing Corporal 'Tug' Wilson at one stage to shout over his radio, 'I'm the point section of this cordon and I'm still getting RPG'd, can I please have some sort of fire support?' Unfortunately his position was 'danger-close' to the enemy firing point, so 'one of the lads crawled up with a 51mm mortar tube to help him out'.

At 0230 Whiskey Company were told to 'advance to contact', as enemy had been spotted on the extraction route, and they, along with Juliet Company, were to clear it. The company walked back through the village, meeting no resistance as they did so, until they came to a square, a walled-in marketplace in which the animals were gathered all day, where they waited to oversee the convoy on its way home. It wasn't the most pleasant of nights, 'sitting around a marketplace that smelt fucking awful, with the lads shouting for people to stay away – until there was a burst of machine gun fire'.

Whiskey Company's job now was to secure one of the two bridges that crossed the river in Gereshk, with Juliet Company taking the second. 'The convoy approached and disappeared to FOB Price, and I thought, "Now let's get the fuck out of here." We were told to follow a path that hand-rails the river, a good linear feature that could be used for enemy mortars against us – eventually we got away from the place via the paths that the locals were now using to make their way to the river for a wash. We were listening to the call for morning prayers, when people appeared, with more weird and wonderful dogs. Shepherds with goats were squatting down staring at us marines walking past,' recalled Stubbings.

It was a long patrol – nearly thirty-six hours – but it had been successful and while they had taken two casualties, they had inflicted many more. Ewen Murchison, like his marines, was impressed by the bravery of the lads from 28 Regiment, Royal Engineers, who had physically built the checkpoints. 'Their task seemed to expand throughout the day, but they kept on: they just continued to dig with mortar rounds landing within metres of them. At one point I spoke to their company commander and suggested they take a half-hour break as the mortars would stop in a minute, but his reply was,

"No! Every minute we don't work we are keeping you guys on the cordon a minute longer." Amazing bravery.'

Back at base, Ewen Murchison was able to look at it as a whole: '*Slate* was a unit-level operation to establish a security ring to the north, north-east and south of Gereshk, culminating in a pretty spectacular baptism of fire for twenty-four hours. It was the first time we had Vikings in a forward deployed cordon. We had to go through the town to reach the checkpoints, which itself posed an IED threat from those Taliban in town; not many but significant if they wanted to make life difficult. Once the inner cordon was established I formed a mobile outer cordon with the Vikings and decided to push out into the green zone and the Helmand River flood plain: a difficult area but Viking is an ideal vehicle for it: first time they had been used on anything other than a convoy operation.'

The Vikings that had arrived for Kilo Company's MOG operations had not been greeted with universal enthusiasm when they were first paraded in front of the marines.

Every Royal Marine believes himself to be a heavily armed commando soldier who relies on intelligent fieldcraft, cunning, speed, surprise, subterfuge, guile, superior training, supreme physical fitness and endurance to take the fight to the enemy. If he had wanted to go to war with armoured vehicles – inside which he could gain no situational awareness as he approached his objective – he would have joined some other branch of Britain's Armed Forces. Simply, the Viking was not for him.

By the tour's end, opinions were to change, even among the most Luddite, and as one of 3 Commando Brigade remarked, 'The most hardened marine can become quite armour-friendly when the RPGs are flying around.' With experience and a growing understanding of the Viking's capabilities – and its limitations – it soon endeared itself to the commando brigade and was, truly, to become not only a life-saver but a battle-winning asset.

Captain Chris Witts, Royal Marines, commander of the 1st

Armoured Support Troop, was, hardly surprisingly, an enthusiast: 'The Royal Marines Armoured Support Company formed in the United Kingdom with brand-new vehicles, operated by brand-new crews manning a brand-new company ready to deploy on active service.' At the beginning of August 2006, the Vikings were loaded into a merchant ship at the military port of Marchwood, Southampton, and sailed to Karachi, from where they were transported across Pakistan via 'jingly' trucks to Kandahar. As a rudimentary camouflage all were covered with tarpaulins; amazingly, all thirty-three turned up.

There is no doubt that the Taliban could have hijacked the convoy. Much of the heavy equipment destined for the United Kingdom Task Force in Regional Command South is brought to Kandahar in convoys of 'jingly' trucks, but it is equally clear that money changes hands to prevent any obstruction to this weakest of links in the logistics train.

Now the two Armoured Support Troops under the overall command of Major Jez Hermer had six weeks at Kandahar to prepare their vehicles to an operational standard. As much of the loose equipment, electronic enhancements and armour was to be airlifted in, the men of the Armoured Support Company spent hours each day at the airhead puzzling out what had not been delivered, what had been delivered and then how to fit it and operate it: as an example, all the thermal cameras, which were new to the drivers, arrived without instructions. The challenge with these and other specialist items was to work out how to drive with them before making out a training programme. Conditions at Kandahar were described as 'chaotic, with no training area, so we could only drive round the perimeter track and not above 20mph; neither was it a particularly safe place for, despite the RAF regiment's "defensive patrols", we were always being mortared'.

With the Vikings in theatre their crews could now marry up their steeds with a number of urgent operational requirements, such as the day/night-vision system which was to prove invaluable:

a thermal camera mounted on the outside is linked to a liquid crystal display screen pulled down over the driver's window so that he drives on a thermal image.

The Viking is a vital mobility asset but it is designed to save lives; yet, although immune to small arms, it needed further enhancement against anti-tank missiles and, especially relevant in Helmand, the ever-present RPG. As the bar armour arrived by air in dribs and drabs it was fitted in stages until, at last, all were fully kitted for action.

Ready for operations, it was decided to 'brigade' the Armoured Support Company's two troops and farm them out on an 'as requested' basis with the brigade operations officer, Major Ollie Lee, tasking them once a bid was received from a sub-unit or Commando.

Chris Witts recalls his surprise at this change of plan, as he had understood that he would be working solely with 42 Commando, who would take all thirty-three vehicles with them, seven vehicles to each company with two crew per vehicle. 'The brigade commander now asked Major Hermer to present a different plan for distributing the vehicles, so we split the team in half. Two troops of thirteen vehicles, each commanded by a warrant officer or captain, who would then work like a helicopter squadron. We would turn up, get briefed, ask the lads to jump in and then deliver them on target. We would organise the fire support and offer casualty evacuation and at the end of an operation pick the lads up and bring them back: a battlefield taxi, although I don't like the term. We would have seven reserve vehicles which would allow us to run the R-and-R plot and retain few spares. We were an "on call" mobility asset.'

The Viking was designed to carry 1,100 kilos in the back, a weight allowance soon taken up by eight Royal Marines who, with full kit, weigh 960 kilos, leaving 140 spare kilos for extra ammunition, water and food. The designed, all-up, weight had begun as nine and a half tons, but increased to 12.9 tons: a heavier weight no longer cleared for amphibious work. Without extra fuel, the

Viking's range was about 150 kilometres. 'Most did not appreciate the weight problem to begin with. All they wanted to do was just fill it up with marines and then add a ton of ammo.'

Despite his undoubted enthusiasm for, and knowledge of, the Viking, Witts retained doubts: 'We didn't know how we were going to be used. All we did know was that we had to be operational by 1 November.'

Through hard work and good leadership, on 1 November the two troops of Vikings were declared operational and ready for the move to Camp Bastion. The direct route from Kandahar is down Highway One, a prospect with which Witts was not happy. 'As so many people told us it was the route of doom and gloom, we studied tracks through the desert, but in the end decided to use the road in two groups and just drove the 180 kilometres in about eight hours with various planned stops and air support from Lynx helicopters.

'Nobody molested us. The only problem was an escorting Danish vehicle which threw a track in the centre of Kandahar, about the one place you do not want to have a breakdown. But the Danes were brilliant and were towing the vehicle in about forty-five minutes.

'Our first task was to support a convoy moving elements of the Commando Logistic Regiment to Now Zad by providing flank, forward and rear protection.

'Suddenly we were being tasked properly. To begin with no one was quite sure who commanded these things. Three vehicles form a section with a Royal Marines corporal commanding; each section can lift a troop and the troop commander sits with the "armoured support" corporal who is his adviser.' To a Royal Marine this is similar to the relationship between the corporal coxswain of a landing craft or a corporal in command of a section of rigid raiding craft; thus the command and control of the Vikings did not, in the end, come as quite such a novel idea nor one to be treated with as much caution as at first feared: in command and control terms, the corps had been here before.

The Vikings proved their worth to the marines at Gereshk, where they were able to transport men quickly and safely to – and from – *Operation Slate*. The protection they offered was vital, for over the time that the men were in the open, about eight hours, more than a hundred mortar rounds were dropped on them.

Some jobs, though, were easier in the lighter, more mobile vehicles. Once the men were in position, Marine Mark Farr, the Quartermaster's assistant, drove round in an open-top Pinzgauer distributing sandbags to the forward troops until a mortar bomb explosion toppled his vehicle into three feet of water at the bottom of an irrigation ditch, with him beneath it. Remarkably, he was quickly rescued by the company QM, Colour Sergeant Carl Tatton, who jumped into the ditch to assess immediately that if they didn't get the truck off Farr, then the young man would drown. Tatton grabbed the roll bar, lifting the truck enough so that Farr – uninjured – could stick his arms out to be pulled free by fellow marines.

A few arrests were made during *Operation Slate 1*. At one end of the cordon the marines and their ANP colleagues saw men in traditional clothing walking past. They were stopped, questioned and invited to turn out their pockets, revealing Pakistani money in their wallets. This could have indicated they were Taliban, or at least had Taliban connections, for having the money meant they were regularly crossing the border. In itself, unfortunately, it proved nothing.

Perhaps the men were out spotting the positions and equipment of the Royal Marines because, when the Vikings were returning after the operation, they ran into a well-coordinated ambush, when one Viking was disabled after an RPG round struck the hydraulic cable that links the two cabs. Ewen Murchison was the officer in command: 'The vehicle – the first of a few – had to be denied to the enemy after the crew and passengers were rescued from the still-active killing area by another vehicle with Corporal Ashley Oates in charge of the evacuation. He was to be Mentioned in Despatches for his bravery.'

The loss of the first Viking was a blow for it showed, as expected, that it did indeed have an Achilles heel; the umbilical cord and linkage between the two cabs were unprotected due to a lack of funds.

At the time the Viking had been transporting men of 2 Troop tasked to climb up Hill 852, from where the Taliban had been shooting down on to the Royal Engineers. The Apaches had made a sweep of the hill but there were still 'black shapes' visible, so the men were sent to see what they could find. The first section was despatched ahead, to clear the hilltop and, once up there, settled into firing positions. Among the men was Marine 'Bungy' Williams, who had joined the corps from St Vincent in the West Indies. Williams is blessed with remarkable eyesight, as his troop sergeant Nigel Quarman explains. 'He is a modest guy and won't tell you that he has this extraordinary ability to see things with the naked eye, some kilometres away. He would pick out mortar base-plates that were firing at us that we couldn't see through our binoculars.'

It was Marine Williams who first spotted what the 'black shapes' were, and they weren't the Taliban. 'Our job was to get to the top to observe and bring down supporting fire, as we were being mortared down by the dam. The hill was our destination, so we continued on to the high ground in order to cover the Royal Engineers doing their thing. We went cross-country and through very tight places until, halfway there, we had to take cover from some Taliban coming down towards us. A small ambush was prepared. The Apache were firing us in and as we could see these black shapes – everyone thought that was what the Apache was firing at but no! It was camels.'

Once they reached the top they understood why the camels had been running away, for the Apaches had caused terrible damage to the Taliban positions, churning the ground up to resemble a ploughed field, flinging solid material everywhere. Sergeant Quarman was in the section that followed the first men to the top. 'As we started to move forward we found what we thought was a dead Taliban who had been hit by an Apache 20mm cannon. But as he had mortar rounds underneath we weren't too sure if he had booby-trapped himself or if his "friends" had done so. As we were about to confirm that he was no longer a threat he moved a hand – so was killed in self-defence.

'Once we were established the Taliban began mortaring us. Quite accurately. We decided it was not sensible to stay unless we could silence their mortars so we were called back to carry out an assault and thought that was a better use of our skills rather than remaining as a sitting target. But on the way back we were ambushed by about twenty Taliban who fired eight RPGs, one of which hit a Viking on its umbilical because we did not have the titanium sleeves that we had asked for as an urgent operational requirement. We were about to cross the Helmand River again via shallow water and a sort of sand-bank which was only about thigh-deep in places and through very thick vegetation, when a second ambush hit Lance Corporal Jim Wright and Mike Gregson's vehicles.'

Jim Wright, a holder of the King's Badge, who had joined Juliet Company late in 2005 straight from training, continues: 'As we broke off the feature we were going through short, tufty grass towards the riverbed. They were waiting for us to get to the river, which is when they brought down the RPGs. Our waggon was hit with a massive strike on the rear cab. The RPG glanced off the bar armour before hitting the linking connections, drives and hydraulics. The Viking just managed to crawl out of the killing area into a little bit of cover by the riverbank, but no further than that as it was pissing out fuel and hydraulic fluid.

'We de-bussed into anti-ambush positions. Seeing our problem, the other waggon's passengers de-bussed as well to cover us. This was all taking time, but we were all right as the cover was quite thick and the machine gunner on the crippled Viking was engaging targets all around us although it was quite tight in. The other lads could see the enemy better and while they dealt with that problem we were busy stripping out our kit. We knew there was no way we could recover the Viking in that area and while in contact, so we threw grenades in and also hit it with phosphorus but we were wading in fuel, which was not a good thing.'

Another member of Juliet Company, Mike Gregson, finishes the tale. 'We stripped out the kit, set fire to the Viking and tried to use the ILAWs – an 84mm unguided anti-armour rocket launcher that is designed to be carried by a single soldier; the one-shot launcher tube is disposed of after the rocket has been fired – but they just bounced off, which was quite impressive from the Viking's point of view. The other Vikings moved out of the killing area to regroup where Corporal Ash Oates and the troop sergeant planned to return for Corporal Lee "Kibbler" Matthews's section. Moving through thigh-deep water and dense vegetation, they found Kibbler's lads and the crew defending the area while trying to strip the waggon.

'After breaking contact with the enemy we arrived back at the dam, in just one Viking with lads clinging to the outside, to find the Royal Engineers still hard at work. We were re-tasked to create defended positions around the outer cordon but we had no kit as our Bergens, slung on the sides of the Viking, had been destroyed, as they often were, by enemy fire and RPGs. As we were being mortared again we set off quickly for our new task, but we now didn't have any food either so the other lads shared half their twenty-four-hour ration packs.'

For Colonel Matt Holmes it had been a good day's work: 'We built three permanent checkpoints in the immediate area of the dam, which was one of the objectives the Taliban kept attacking. We'd been near the Gereshk dam for no more than twenty minutes when we came in contact and remained in contact for the next twenty-four hours or so. We knew the Taliban would attack from all sides, which they did, as there is no clear front line of troops, or FLOT – they would always try to encircle us then pop up anywhere. After an hour or so, I had a big grin on my face because everything was working – the train set was working – everyone was doing their job and the day got better as we went on. For instance, I asked the officer commanding my mortars, Captain Rob Thorpe, to tell me if it was his or theirs that were firing so we didn't have to keep taking cover every minute. Also every man was getting his compass out and trying to pinpoint the enemy as the enemy would fire two or three rounds, then move position – like all well-trained infantrymen.'

By the morning and as the ANP began to take charge of their new checkpoints, the operation drew to a satisfactory close and, for many, a yomp back across the desert to FOB Price where the company sergeant major, Scott Burney, was reunited with his beloved quad bike, alarmed, though, to find it full of holes, having been caught in various crossfires.

There was still the question of the disabled Viking to deal with. An SNCO from the Armoured Support Group announced that it would take a 2,000-pound bomb to destroy the Viking, so a Dutch F18 Hornet, patrolling overhead, was tasked with denying the vehicle to the enemy: he only had a 500-pound bomb which, landing close by, simply knocked the vehicle on to its side. The original request for a 2,000-pounder was reinstated and later that day a Lynx helicopter flew past the site to report that the Taliban were clambering over it as it lay smoking. About two months later a Desert Hawk's video camera confirmed what all suspected would occur: the Taliban had dragged it away altogether for their own purposes.

Short of a suicide mission – and there were spares anyway – the Viking was doomed. Ewen Murchison, whose decision to destroy the vehicle it had been, reported: 'The Viking needed to be denied so it was pleasing that higher command made it quite clear that no lives were to be hazarded by trying to rescue a piece of hardware.

If the only way forward was to deny it to the enemy and allow the men to escape the area, then that decision would be fully supported. In the end I made the decision to bomb it and not the brave corporal, who had enough on his plate without having to worry about making that sort of decision: to destroy a million pounds of gear.

'To me and Juliet Company this operation was an important first step, as it was ISAF putting a security framework on Gereshk itself, facing down the Taliban and being seen to stand firm in support of the local police. The locals were appreciative, but very quickly the Taliban, inevitably, tried to counter what we were doing and, at night, began to attack the checkpoints we had established. The police were not trained for such events and needed a bit of backbone, which we could supply by regular deployments from FOB Price in the dark and at the rush if we had any sort of intelligence: more than an Afghan policeman ringing me up saying he was about to be attacked by five hundred people. We never actually manned the checkpoints with them but I would often have my men sitting to the rear on high ground, just to give the ANP the happy feeling that we were there and that if they did get attacked we could physically support them with air and artillery.'

42 Commando's commanding officer was also pleased but wanted more. 'We launched *Operation Slate 2* a few days later. I still had field engineers under command and we knew we had walloped the Taliban that day and pushed him back. So, to continue the exploitation phase of the area, as it were, I grabbed Mike Company Group and the engineers again and they deployed along Highway One to the eastern side of the river. There they put in a very substantial vehicle checkpoint, reinforced existing defences, built new ones and trained the ANP how to conduct vehicle searches: all led by Major Martin Collin, who concentrated on this checkpoint as it covered a bridge over the river. This little operation had a significant impact, as it was a very high-profile

checkpoint right on Highway One and one with a far higher profile than the original meagre affair that had been there. The issue of course were the ANP whom we left to man it, for they would often rob the locals by demanding large and illegal "tolls".'

Sergeant Jay Layton of Mike Company's Fire Support Team was called out for *Operation Slate 2*. 'I moved my FST troop on to a piece of high ground from where we could overlook the river and surrounding fields.' It turned out that they were in old Russian defensive positions. The rest of the company remained out on the main road, setting up temporary vehicle checkpoints, slowing down traffic and searching vehicles. 'We were there for about five hours and then suddenly about one kilometre away there was one hell of an explosion and off to our left we had some of the 29 Commando Regiment guys calling in artillery, Apache and "fast air". We all shouted, "Take cover!" but didn't have much, as we were still building our defences. We thought it quite amusing as we were all standing there with our picks in our hands until someone said, "Why did you say take cover? It was over one kilometre away." Then we heard a thump in the distance followed by the whistle. Instant reaction: Where's me helmet? Where's me Osprey? We were getting our heavy weapons ready, including Javelin, and then this round comes in about 200 metres in front of us. Big cloud of dust and shrapnel whizzing over the top and then a second thump and everyone is shouting, "Incoming." The next one fell behind by about a hundred metres, so they were bracketing us. Then we heard a third thump and more whistling and this one landed in between our positions, directly on top of the hill. They had got our range fair and square. Must have been controlled by a professional Pakistani. Then they fired a further three rounds creeping towards the Royal Engineers in their JCBs but they just kept on going with the rounds landing in among them – fantastic fellows. I watched them being sprayed with earth as they continued to erect and then fill the lines of Hesco Bastion containers.'

From his position on the hill Layton tried to identify the enemy base-plate positions, having assessed that they were probably 120mm mortars – the largest yet. But it was impossible, so he called for 'fast air' to fly over to try and locate them. After a further forty-five minutes and three more bombs they managed to target the base-plates. 'From an Ops One point of view that was it and we went back to Camp Bastion.'

It was during the two *Operation Slate*s that the penny started to drop about the use – and value – of the Viking. They lifted the troops in and dropped them before supplying fire support and carrying out remote casualty evacuation. The official view was that 'without influencing the ground battle they were independent and able to move around on request'.

This remarkable vehicle was fast becoming fundamental to the brigade's success. 'Viking was undoubtedly a force multiplier that contributed significantly to our operational effectiveness,' states Ewen Murchison. 'As a patrol vehicle, not as an infantry fighting vehicle, it was extremely effective at crossing difficult ground, allowing the company group to dominate the terrain and manoeuvre to deliver rapid effect across the area of operations, achieve surprise and thus contribute to the disrupt effect. As well as the manoeuvre capability, the psychological effect should not be underestimated: there was much ICOM chatter about "tanks" in the area; however, for outreach tasks and village assessments the Viking can have the undesired effect of intimidating the locals.

'The enhanced protection, coupled with the additional firepower delivered by the vehicle-mounted GPMG – giving a total of twenty-three GPMGs in the company group – contributed to both our physical and moral components. Even the most ardent Viking sceptics were won over after our large contact in the green zone at Habibbollah Kalay in January, during which five of the Vikings were hit by RPGs.

'The Viking also offers a very robust communications

infrastructure for company-level operations, with each vehicle having a full electronic counter-measures suite. The only constraint was the reduced situational awareness,' with Murchison comparing it unfavourably to the visibility available from the open Pinzgauer troop carriers, which they used in conjunction with the more heavily armed Land Rovers. 'After losing a Viking on *Operation Slate*, we realised that the two should always be employed in tandem,' Murchison concluded.

Murchison's views were similar to those held throughout the brigade: 'It is an outstanding piece of equipment, fit for purpose and well crewed by the dedicated and enthusiastic men of the Armoured Support Branch of the Royal Marines, who gained in experience rapidly throughout the tour and worked tirelessly to refine SOPs to meet the changeable operational environment.'

Elsewhere across Helmand Province life, too, was processing at a hectic rate. At Sangin, Lima Company with Major Steve Liddle and CSM Billy Bowker were coming to the end of their planned two months in the District Centre and preparing to hand over to B Company the Light Infantry, the theatre reserve troops. By coincidence Liddle had served with the Battalion in Bosnia and knew a number of the incoming soldiers.

The ceasefire in Sangin was holding, with Liddle conducting regular *shura* every two or three days in his HQ. At each meeting he would reassure the elders that the British would not be leaving while the place was secure. In return, he relied on them to explain how things were going, but it struck him that they were more concerned not with who held the power in the area but rather with who would be the next chief of police. He respected the elders but was convinced that the real decision makers were still the Taliban. He was equally convinced that some of those at these *shura* were Taliban – 'who definitely wanted us out and saw us as the cause of all the earlier trouble'.

He, too, found the local attitude polite but puzzling, for any suggestion of help was always refused. 'In November they lost a truck in the river and we offered to help them recover it. It was something positive but they did not want our assistance at that time. So we left at the end of November and moved to Camp Bastion as "Ops One Company" till the end of January, when we took over the District Centre in Now Zad.'

It was clear that regular face-to-face meetings with the locals, whether just in the street or at the *shura*, while not immediately having an effect, were vital to the Coalition Forces in hastening the advance of peace. In Gereshk, an area that up till now had been more confrontational than it was currently, it was having the men visible that made the difference. The twin strategy was to use night patrols, and the by now ever-present Vikings with their night-vision capabilities, to comb the area in the hours of darkness, or to conduct 'robust' actions against the Taliban. Then during the day the men would patrol the town, but this time in a more approachable style, wearing their green berets and not helmets to emphasise this.

Under the Paras, this hadn't happened, but that had not been their fault – it was a factor of the enhanced capabilities the Vikings brought with them, as well as allocation of men to this troubled area. 'Our predecessors had not been able to invest in Gereshk as they were focused elsewhere,' said Ewen Murchison, 'but it has to be fair to say that the Para officer I took over from knew that his credibility with the locals was lacking for he had been unable to do anything – through no fault of his own, of course. We now had a footprint of credibility in the town: they could see we were taking on the Taliban and at the same time helping the police and the army.'

With the feeling that things were settling around them, Murchison was able to expand the strength, style and length of his operations, with the aim of pushing the Taliban further back from the town. Not all of these operations were run strictly by the book, for in much the same way that the men were learning the new art of 'advancing to ambush' in order to draw out the Taliban, with the mobility he now possessed Murchison could begin his programme of domination throughout his region. 'It was the ideal time to start Combat Recce Patrols for the very practical reason that if you go out on a deliberate operation the level of control and the amount of paperwork involved in advance are prohibitive; but if you go into an area just to feel your way around, to see what would happen, and *then* became engaged – it is not a fighting patrol but a genuine recce!

'You can do that off your own bat and quickly but you would not get pre-planned resources – fire support, air and so on – but if you actually got into a TIC – troops-in-contact – you would soon get help. So, after a couple of months of shoring up the town those recces became an enduring line of operations; and all the while pushing the envelope with the aim of establishing a buffer zone of about five kilometres outside Gereshk – mainly to the east – as that would effectively take the town centre out of range of Taliban fire. This involved dominating various spot heights such as Hills 836, 828 and 852. Ideally I wanted to hold this key terrain, not permanently as we did not have enough troops, but we needed to be able to use these areas as and when we wanted. He who controlled the heights controlled the valley and town.'

The patrols would head out into the desert, or skirt along the edges of the green zone. In order to avoid being ambushed too easily, they'd vary the pattern, setting out at different times of the day or night, never going in the same direction twice. Sometimes they'd set off in vehicles; sometimes they'd head out in helicopters, 'dusting down' and re-embarking before the Taliban could track them. On one patrol they waded up the river, waist-deep, the company's mountain leader up front guiding them. Each patrol allowed them to build up a picture of the area, to know more about the way of life of the people there, and of course to find the areas where they would most likely come into contact with the enemy. The men would take cameras with them and snap off pictures of the cave-like houses built into the rocks at the side of the river, now mostly abandoned as the former inhabitants tried to avoid getting caught up in the fighting.

'Every time we entered the green zone I would have a contact, simple as that,' said Ewen Murchison. 'As it is very close country, like the Normandy *bocage*, with deep irrigation ditches, hedgerows, very narrow paths which, if you manage to get a vehicle down completely channels it – can't turn around. Severely restricted for a vehicle and individual mobility.'

Juliet Company were not alone, out there in the desert. During November the men of C Squadron the Light Dragoons under Major Ben Warrack were deployed in their light tanks as part of the Mobile Operation Group (South) or MOG(S) that formed the *Operation Glacier* series. The units, from very different parts of the Armed Forces, might on other occasions have expected to encounter difficulties in working together, but not here, not now.

'Chris Ordway of the Brigade Reconnaissance Force – BRF – was there a month before us with his WMIKs and knew the ground well already,' says Warrack. 'But they were sceptical of us to begin with as they are serious specialists and cooler than cool: sort of Long Range Desert Group, the Royal Marines' light recce specialists and almost all trained as special forces or mountain leaders. When they were told they would be working with a bunch of cavalrymen from the Light Dragoons I think they were appalled. However, after a couple of decent scraps they realised that we had more firepower, more capability, than they had and we could do everything at night that they couldn't do. Suddenly they wanted us on board and all mutual suspicions that we were "fat cavalrymen" and they were bearded Walter Mittys were lost. We formed a highly synergistic and effective relationship for the rest of the tour. This was at every level and it worked because we supported them and they would be very good at doing close target recces and discrete strike operations but, until we appeared, they had no support force to fire them in or fire them out when things went wrong.

'We could put a huge armoured umbrella over them with our many weapon systems by day and night. For instance, we had sixteen gun vehicles plus a support troop with dismountable GPMGs and sniper rifles, which was more of a Close Target Recce Troop than a gun troop. So we quickly became a manoeuvre support capability for the BRF, able to escort them into an area where they were conducting a recce. We would then melt back into the desert and stay hidden until it "went noisy" or something went wrong, when we would appear over the hill with a squadron of tanks and provide hard support for their extraction.

'Down by the fishhook – way south of Garmsir – we were out of helicopter range for casualties but negotiated a deal where we would take a medic forward and then relay casualties back. When the brigadier found out that we were operating around the fish-hook he was initially furious but then impressed that we got all the way without him knowing.

'The known route for Pakistanis and Afghans working with the Taliban started around Bahramchah on the southern border, from where they would make their way north to the fishhook and the green zone. Then they would disappear, for the Taliban – not us – held the ground.

'Our first big scrap was at Kattaka on 9 November, which was not an official operation but an impromptu, unplanned affair. In our mission statement we were proving river crossing points in the south of Helmand. We knew something was going to happen but we didn't expect to be fully ambushed by dozens of Taliban. They waited and waited and waited until we had completed the river crossing from west to east and were way across the green zone. I was a bit concerned as we were stretched over about three kilo-metres and without clearance to take the squadron over and establish ourselves in the eastern desert. I decided to return and as soon as the order went out we were ambushed just as we were strung out as far as possible. The enemy were very professional, bloody good and bloody brave. They will keep going and going and don't mind dying.

'We fought for about five hours and at one point were surrounded on three sides with the desert on the fourth side – at our back. But we couldn't just turn round and go hell for leather as vehicles were breaking down and had to be recovered under fire. One even got bogged in in the river and that had to be evacuated. The whole thing went on and on and on with small arms, RPG, heavy mortars, heavy Soviet machine guns – the lot.

'Before Christmas the BRF were east of the river and we were west of the river, but after *Operation Baaz Tsuka* in Kandahar we went to the CO of the IX Group and said we thought we could work better together than separately as we had such different but complementary capabilities. On our return from Kandahar that is what happened, to our mutual success, as our joint tasks were to look for Taliban routes into the country and to identify potential projects for investment and reconstruction, such as schools. There was an American-built school down at the fishhook but any child that had been to it had been killed by the Taliban, who would not let any child use it, so that wasn't working at all.

'Other tasks were to make feints and demonstrations when "discrete operations" were being conducted but, apart from causing a bit of a diversion while they were conducting an operation, we never played with either the SAS or the SBS.

'On patrol we had no white light at night, no hot food at night, no washing or shaving for a month and pretty ghastly rations, but our Quartermaster did manage to bring the odd orange to supplement the compo. We never killed a goat – we never saw a goat where we were. In Iraq the Arabs would kill sheep for us on arrival and present the head – which can get a bit too much.'

4

DECEMBER: INFORMATION EXPLOITATION

'As I ran I could see the enemy rounds hitting the compound walls all around me, spraying loose stones and mud towards my feet. I vividly recall a burst of machine-gun fire spurting along the ground towards my legs. I jumped into cover as the rounds came up towards my thighs, narrowly missing my groin area. As I landed I checked if everything was still there before giving Gav covering fire for him to move.'

At the start of December, in Now Zad, Neil Sutherland's reduced-strength Kilo Company was looking forward to welcoming back Luke Kenny's troop from their stint at Kajaki. As well as returning to the company, 5 Troop would help resupply the base. After five days' rest and briefings at Camp Bastion, Kenny's men were ready to join their comrades, but a shortage of serviceable Chinooks meant that what might have been a fifty-kilometre flight would instead be a hundred-kilometre off-road journey in a ten-vehicle convoy of Viking, WMIC, Pinzgauer and Danish armoured vehicles.

Not surprisingly, no one was enthusiastic about this, for, no matter how circuitous the chosen route, vehicles and their occupants would be vulnerable to attack at river fords, wadi crossings, small villages and almost anywhere else that the Taliban, or the Russians before them, might have mined. Every gully or sandhill could conceal an enemy RPG or 107 rocket launch site. So night and day anti-ambush drills and mine strike procedures were repractised.

Al Hewett, appointed a Viking commander, was to use his section as a reconnaissance unit, starting out at the rear of the convoy but quickly moving up the flanks, off the forward edge. There were two reasons for this manoeuvre: somebody had to pass back warning of a possible build-up of civilians ahead, for the differences between their dress and Taliban fighting uniform were indistinguishable; and the convoy needed a 'route finder'. As winter approached, the wadi beds were filling with run-off from the mountains, offering the real and often-experienced danger of flash floods; even with slow-running streams, crossings were slow and difficult, making for ideal ambush conditions. Added to which the Taliban were known to lay anti-tank mines, indeterminately, to halt the unwary; they were also experienced in swiftly co-ordinating and moving their forces to cover any mine strike. Al Hewett's section travelling off on a flank, as their highly visible convoy approached through its own dust trail, increased the possibility of the lead Vikings catching everything first.

Despite the inevitable vehicle breakdowns, and a tense transit of some unmarked minefields, the troop's ten-hour drive to Now Zad was, in the end, uneventful. The views en route through the desert and lower hills were 'stunning, untouched by Western "sophistication", with the surrounding mountains and local scenery breathtaking. When we passed through the small villages, the locals, who had no idea of Western technology or, indeed, of Westerners, just looked on in astonishment as we drove by.'

The arrival of the convoy was greeted warmly at Now Zad, and the stores quickly unloaded directly into the District Centre, while 5 Troop was escorted through the battle-scarred, sandbagged and well-defended compound into the company briefing room for an immediate update on the current situation. Speed was vital as the enemy had a habit of greeting newcomers with a hail of rockets and small-arms fire: they needed to know where their stand-to positions were, and the individual procedures and actions to be taken.

The District Centre held some surprises for the new men from the comparative warmth of Camp Bastion and the desert sun. It was colder here and while their new home, an ex-Afghan police HQ, had sleeping areas indoors – in the cells to the rear – every wall and roof was pockmarked with bullet strikes and the visual evidence of exploded hand grenades, RPGs and close-quarter fighting. They had to get used to the uninvited guests that crept into their sleeping quarters, and to the rationing of water for almost every activity. Truly, 'Now Zad was the worst of the lot,' as Neil Sutherland said. And the set-up at the DC had to be endured by Kilo Company for thirteen weeks – almost a prison sentence in itself.

Kilo Company was still 'enjoying' the precarious ceasefire that Sutherland had negotiated and which allowed him to conduct a limited patrol programme. Incoming fire had never stopped but it was a nuisance rather than a serious threat to the fragile peace.

On 8 December, led by their company commander, sixty-five marines from a manoeuvre troop, company HQ and a Fire Support Troop conducted a framework patrol to the north and east of Now Zad. The aim was, as always, to dominate the ground outside the town and by doing so, show the Taliban that they could be engaged not just in Now Zad itself – when they chose to attack – but in their own enclaves among the woods and rough compounds in that area.

All patrols set out from the base and carried out their task on foot; for support, a Pinzgauer followed the men, while the heavy guns of the fire group sat off to a flank in WMIKs. Both of these elements were able to supply flanking fire, local protection and a permanent over-watch. Sutherland well knew that, ceasefire or no ceasefire, the Taliban would try to cut him off as they had done a few times previously, but he and his junior commanders took great care to choose and then protect extraction routes and casevac plans. Before any patrol set out, one question that always had to be answered was, 'If we are far out, can we get a casualty back?'

The Royal Marines couldn't move from the District Centre without being closely observed; the Taliban had spotters able to track and communicate with the main force in the scrub and hills behind. Listening in to their communications – the ICOM chatter that gave the British the edge in keeping themselves informed about the enemy – let Sutherland's men know how many Taliban they were likely to be facing, roughly where they were, and what support they were expecting. On this day, after the Taliban had watched the marines leave the District Centre, they had split into groups along the edge of the desert scrub and wooded area, probably about forty or so of them, while a number of four-wheel-drive pick-ups were moving up the wadi from the south to provide support. It was clear from all that was overheard that the enemy thought the marines were moving in on their positions in the north-east of Now Zad. The patrol ran into fierce fire and had to extract and return to the DC. 'Although we were happy to fight them, the terrain was too close country for one under-strength rifle troop to deal with. Their tunnel system didn't help us either.'

This was the area to which the majority of the Taliban had moved for the duration of the ceasefire. A maze of tracks, tunnels, irrigation and drainage ditches through scrub and thick, low, leafy trees, that offered perfect cover, it was ideal from the Taliban's point of view – and a pain for the British.

Corporal Al Hewett was involved in an action that day which, when added to further work at Kajaki in the New Year, was to win him the Military Cross. The information and briefing that 1 Section attended lasted an hour and ended with the company commander expressing his view that their patrol wouldn't see much, as the ceasefire was still in place. Patrol start times varied, often because the commander wanted to wait until air support was available, as a precaution; on this particular day, the patrol set off early. The air was still crisp and clear, and the heat that would come when the sun was fully up wasn't yet enough to make them sweat just standing there in their kit; that would come later.

The men set off on foot, close enough to support each other but far enough apart so as not to present an easy target. After about one and a half kilometres the men to the front of the patrol spotted a vehicle driving towards them; Afghan villagers, by and large, do not drive large vehicles, so they were immediately suspicious of this one. Permission was given for a hasty vehicle checkpoint to be established, a procedure in which 1 Section was well drilled: Hewett placed his machine gun group in over-watch to cover not only the men who'd stop the vehicle, but also the nearby fields and a straggle of buildings that surrounded the immediate area.

The vehicle moved closer, and the section's interpreter was summoned forward; Al Hewett moved to the centre of the road and, standing in its path, raised his rifle to his shoulder and took aim at the driver. Marine Dan Ravenscroft stood off to the left, ready to step forward to look inside; he would lead the assault if anyone made a suspicious or hostile move. The vehicle slowed to a halt roughly fifteen yards ahead of him, but Hewett wanted to take no chances, so he firmly shouted out that the occupants of the vehicle should all step out. Only the driver did so. Dan Ravenscroft 'satellited' around his corporal and peered inside the vehicle.

'Dan signalled that there were still people in the vehicle. I asked the driver again but he insisted that there was no one else so I ordered Dan to remove the remaining three passengers. As they came out I looked at the driver. He knew he had lied and was beginning to flap. Afghan men regard women as worthless and he obviously hadn't give them a single thought.'

The driver was, the interpreter said, en route for Lashkar Gah; he had no links with the Taliban and had been bullied into avoiding British forces. Now he was scared of the reprisals, for being seen talking at a checkpoint was not good for his future. Reminded that far greater reprisals were in store if he failed to tell Corporal Hewett where his enemy were, the innocent Afghan paused – very briefly – then pointed excitedly, in every direction, explaining that they were all over the place. He wanted to get this unpleasant episode over with as quickly as possible. With time running out and the troop commander demanding constant updates by radio, Corporal Hewett was obliged to move the vehicle on.

Near the northernmost checkpoint an increase in the movement of civilians had been spotted that, with the associated ICOM chatter, indicated the Taliban were now 'on to' the patrol and beginning to set up their defences. The problem that faced Sutherland was the possibility of enemy in large numbers, in defence of and probably astride the planned route. It was important, though, not just to test the effectiveness of the ceasefire but also to show the Taliban there was nowhere the marines wouldn't go, so the decision was made to carry on moving east and then back south towards the DC through friendlier areas.

Moving in open formation, the marines found the ground opened out into large fields to the front and the south, while to the north a series of compounds, two to three hundred metres away, dominated their track. The section was no longer in the cover of alleyways and buildings and suddenly, in Hewett's view, it was overexposed. He ordered his men to adopt an even more open, larger, staggered file formation with the gun group at the rear; the patrol was now roughly in the shape of a diamond, ready to swing in any direction should they come in contact: another well-rehearsed drill. Al Hewett pushed Dan Ravenscroft forward as the point man, at the head of the diamond, and told him to head towards the next checkpoint, a small compound nicknamed Taunton; less than two hundred metres ahead.

1 Section's task was to carry out a quick reconnaissance of Taunton, to see whether or not it had been – or currently was – in use by the Taliban. Hewett was acutely aware of the increasing

enemy ICOM chatter as 1 Section moved forward to their objective: whatever could be picked up by the equipment in the operations room in the old Police Compound was being relayed to the commanders on the ground. The remainder of the troop and Coy TAC remained in place to the east, about three hundred metres from Hewett's section, where yet more compounds edged out into the fields.

The patrol halted in the open for one minute as the situation was reassessed, then continued its cautious move forward. By now the ICOM had increased dramatically; the Taliban had spotted the patrol. More worryingly, the observations that could be overheard were alarmingly accurate – they 'could see a single section moving forward' – and were clearly preparing to take on the patrol. 'Sometimes when the enemy know we listen to their radios, they give misleading information,' said Sutherland. 'It's then up to us to decide what is genuine and what is false. This seemed genuine.' Hewett decided that it would not be good 'for his lads to know the full extent of the ICOM chatter as it would only make them tense and that could lead to trigger-happy accidents, although I suppose I could have given them a cut-down version'. In this case he simply relayed the news that 'the enemy had spotted a patrol on the ground, so get ready'.

Scanning the ground immediately ahead for cover if they were to be shot at, Hewett managed to pick out several small, man-made irrigation ditches of varying length that ran away from the compounds; some were no more than the height of a street kerb-stone, while others were rather deeper and thus more attractive as cover. All, however, were discoloured, stained and stank, for they were full of human waste from the Taliban, who used them as lavatories, living as they did out in the open.

The patrol advanced closer to Taunton, moving slowly and ready at any moment to take cover and engage the enemy. When Dan Ravenscroft was about fifty metres away from the compound, meaning that the gun team were nearly ninety metres from it, the silence of the still countryside was broken by the crack of automatic AK machine-gun fire and the *whiz* as 'Afghan bees' zipped close overhead and through the section. Rounds smacked into the sand around them with the ground tearing up in thick patches that flew about. 'First-timers' are often mesmerised by this 'ground lifting', but these marines were not.

'Contact left!' Hewett screamed. The Taliban had engaged 1 Section from a number of well-prepared firing points to their north, some two hundred and fifty metres away. The section did what they'd been trained to do and immediately dropped down, some kneeling, raised their weapons and started to return fire. The noise was deafening as all five men simultaneously poured hundreds of rounds towards the enemy, shouting as they did so, yelling to each other how far away the enemy were and what they could see of them. Dan Ravenscroft, who'd been waiting for this moment, popped an under-slung grenade launcher bomb in the direction of the fire coming their way. Even before it had struck he was reloading immediately with a smoke grenade that, when exploded, gave his section a chance to break towards cover – the uncleared compound, Taunton.

'As Dan moved towards the western edge of the compound to lay down covering fire I gave the order to "break" – meaning to peel into Taunton,' said Hewett. As soon as the smoke bomb landed the men were up and running towards the relative safety of the compound while Al Hewett stood his ground and kept firing to suppress the enemy position while the rest of his section 'fired and manoeuvred' past him and into shelter. Once everyone was inside the compound Hewett called up the DC to give a contact report to the boss. 'At that moment Tom Lockyer – a newly joined marine – pointed out a man with an AK-47 running out of the eastern edge of the compound that we were next to and heading towards a more southern compound we had nicknamed Exeter –

our next checkpoint. Tom dropped him but we couldn't tell if he had been killed or not.'

The men were catching their breath and reloading when, as Hewett says, 'the worst happened': the Taliban, who had waited not only in Exeter to the south but also in the wood line to the east, opened fire on them in their hiding place, the bullets thudding into the soft compound walls – with the wall they were leaning against starting to split apart and crumble. They had to move, and fast, for they were swiftly being surrounded. The best – the only – choice open to them to leave the area and rejoin the remainder of the patrol was along the line they'd come, yet it too was open to enemy fire.

The section leader quickly set up the GPMG so that it could be trained on the heavier Taliban element to the north; and the other men were concentrating their fire on the wood line and on Exeter, not able to pick and choose targets but laying down general fire. While they did so Hewett scanned the ground to see what he could do to get his men out of there and noticed an irrigation ditch running west to a compound known as Bristol, some three hundred and fifty metres away.

Time to go. As one of the marines was heard to say, 'Fuckin' discretion is the better part of fuckin' valour.' Above the noise of battle Al Hewett indicated the route he wanted them to take and shouted at his men, 'Make for the ditch. In turns. Crawl west towards Bristol. I'll cover you.'

One by one the men raced out of the shelter of the compound into the irrigation ditch. Al Hewett stood and fired short bursts to give each man cover as he ran, before it was his turn to race over and dive head-first into the shallow protection the ditch provided. It was worse than he thought it might be, the foul stench causing him to gag, and he had to force himself to crawl through yards and yards of human excrement, all the time keeping his head lowered close to the filth in the ditch while the enemy fire from the north increased as they tried to pin the men down.

Eventually he came to a point in the ditch where the slope rose enough to give cover from the enemy to the north; the men were gathered here, waiting for him, all smeared in the same ordure from the ditch. Unfortunately they were still sufficiently in the open to be taking 'incoming' from the eastern wood line and southern compounds. Hewett looked about him and realised fractionally ahead of the others that their only option was somehow to make it to Bristol, and to do that they would have to cross exposed, coverless ground. 'We all realised that to get there we would have to fire and manoeuvre in the open while fending off attacks from three sides.' And once at Bristol they would need help to get further back, as the Taliban would surely be looking at closing their last exit route off, if they could.

Using the small amount of cover available to him, and shouting to be heard over the shooting going on around him, Hewett contacted the troop commander, to relay a situation report and to request that support be moved up towards the southern edge of Bristol at the same time that a flanking section be moved to rendezvous with him there to help his extraction. This would also mean – he hoped – the suppression of the targets to the east and south. To help his troop commander he passed precise grid references of the enemy's in-depth locations to the north, east and south, enabling – further up the command chain – the company commander to direct fire from the 81mm mortars stationed on ANP hill.

What Corporal Hewett was also told was that, while his section had been 'busy' with their contact, the moment the shooting had begun, Coy TAC had requested emergency close air and Apache support. The aircraft were already closing in to join the fight.

Hewett quickly outlined the situation and his plan to his section. With no eastern entrance to the compound they would still be taking incoming fire from three directions, but he hoped the flanking section would be in place quickly to cover them against the shooting from the east and south. This left the northern sector the only one for them to suppress, but with the men prepared to fight their way to Bristol, Hewett drew a deep breath, then shouted, 'Go! Go! Go!' Needing no further spur the marines jumped from cover, each firing a five-second burst in the direction of the Taliban to the north, hoping to force their heads down for a few moments at least. Marine Ravenscroft again fired a smoke grenade as the men hopped out of the ditch to begin pepper-potting towards Bristol, flinging themselves down into a shallow irrigation ditch as the enemy bullets starting flying among them again. This ditch too was full of human waste, and too shallow for proper cover, so every ten metres or so the men would prop them-selves up on the earthy bank and fire back at the enemy. Over from the east the Taliban started to concentrate the majority of their efforts on this 'new' ditch, letting off bursts from their heavy machine guns and lobbing over RPGs.

Hewett knew their situation was tough: 'The explosions were deafening as the RPGs exploded behind us, mingling with the crack and thump of small arms and ever increasing machine-gun fire. We were well and truly pinned down. I lay flat for a moment while the enemy concentrated an enormous rate of fire on our posi-tion from the east and, for the first time, even contemplated not making it out alive.

At that precise moment an Apache attack helicopter dived down, firing rockets and 30mm cannon into the enemy positions to the east as Kilo Company's own 81mm mortar fire, directed from ANP Hill, started to fall among the enemy's northerly positions.

Scrambling for fifty metres along the shallow irrigation ditch, Corporal Hewett and his men now had no alternative but to move into the open, for it was imperative that they link up with the flank-ing fire team, well positioned along Bristol's southern wall. 'Looking at the lads I knew they were tired and I knew that ammu-nition was low. Quickly I briefed them of my plan to move and link up with the troop; myself and Gav with the GPMG would stay where we were and suppress the positions to the east while 81mm mortar fire missions, along with Apache helicopters, would concentrate their fire to the north. This would, I hoped, assist the extraction of the remainder of the section. The section – minus me and Gav – would link up with the flanking fire team and provide covering fire for us two to then extract.'

It all sounds logical, clear and well thought-out, but this was being conducted under immense pressure, with fire coming in from all around them and only seconds to form, weigh up, and decide upon what actions might – or might not – save the lives of his men. Hewett's decisive actions certainly benefited from his commando training, which now came to the fore.

Hewett forced himself up to stand in the line of fire, shooting even as he brought his gun up to his shoulder. Alongside him Gav weighed in with the GPMG, spraying bullets over the area to their east. The three remaining members of the section – Dan, Tom and Dave – broke cover and ran over to Bristol to join up with the flanking fire team. Al Hewett and Gav dropped down to reload before it was their turn to move – at the moment they did so, enemy fire shot up the ground all around them. 'Winning the firefight with the enemy was extremely hard, as they were in numerous in-depth positions, all coordinating fire on to us. I knew they were still trying to encircle the patrol but, for now, I was happy that the other three lads were with the remainder of the troop. I looked over at Gav and was proud to be with him. He is a strong character and didn't think so much about dying: he just got on with the job and was very calm under fire, almost laid-back. A credit to me and the corps that day, he was continuously laying down suppressing fire on to the enemy positions as I was on the radio, while in turn I gave him suppressing fire so he could clear weapon stoppages.'

Rather than dash across the open ground at the same time, giving the Taliban more to aim at with no one returning fire, the two agreed that they would move off together but that they would stagger their runs forward, giving each other cover as they went. Hewett would go first, then provide cover for Gav in return.

'We looked at each other before moving, then, after one final "Good luck, mate," we were off. I moved, then he moved, constantly covering each other. As I ran I could see the enemy rounds hitting the compound walls all around me, spraying loose stones and mud towards my feet. I vividly recall a burst of machine-gun fire spurting along the ground towards my legs. I jumped into cover as the rounds came up towards my thighs, narrowly missing my groin area. As I landed I checked if everything was still there before giving Gav covering fire for him to move.'

The two continued to manoeuvre towards Bristol until they were near enough for Hewett to dash to safety. 'Approaching the flanking fire team I sprinted to the south side of the compound to link up with the troop, then, just as I did so, the enemy increased the rate of fire in one last effort to take us out.'

This left Hewett's last man exposed, but now he had the flanking team providing covering support, along with the rest of the section urging him on: 'Gav was still ten metres away and as I turned to encourage him under all that enemy fire, the suppressing rounds fell silent due to weapon stoppages. This was not ideal as it left Gav moving in the open unsupported.'

For Corporal Hewett there was only one answer. He stepped back out into the open, moved away from Bristol's wall in order to cover his friend, and, switching his weapon to automatic, swept fire across the nearest enemy position, emptying a whole magazine in the process. Gav ran past him shouting an unintelligible mix of swearing and thanks, before Hewett stepped back: the section was together once more.

Although the teams had now linked up, they were not yet out of contact with the Taliban. To return to the remainder of the troop and the quick reaction force they still had to cover some two hundred metres of open ground towards the compounds where, at the start, the troop and Company TAC HQ had gone firm at the same time that 1 Section was first engaged. This ground was exposed to fire from the north, but already a build-up of fire from the 81mm mortars, together with developing mortar smoke, offered increasing cover for the final move to the remainder of the troop.

The section commander briefed his men for yet another danger-ous dash across exposed and cover-free ground. As they began their latest fire-and-movement manoeuvre towards the troop position, he ensured that he was the last to leave. If casualties were to occur, Hewett wanted to be the first to spot them, enabling him to react to the situation more effectively.

'Moving across the open ground, I could see exhausted men carrying enormous weights, running for cover: this, in my opinion, was where our training paid off. The physical and mental state of my men was far superior to most, and it showed. Despite being in a rather degraded state they performed well.'

At last the 'link-up' was achieved but, with no time for rest or banter, an 'ammunition and casualty report' was demanded along with a detailed check that all 'mission-essential' equipment such as electronic counter-measures equipment and radios had not been lost or damaged by gunfire. The professionals were still at work.

'It was good to see friendly faces, although I will always remem-ber the look that all the lads gave the section as, covered in human waste, we came in through the DC's gates. Dutchy Holland, the quick reaction force commander, came over to shake my hand and to tell me that he had thought all of us were undoubtedly dead.'

In the District Centre there was no time for further pleasantries; only time to drink water, recharge ammunition and prepare for the next mission as, overhead, Apaches, Harriers and mortars contin-ued to pound the enemy, along with the HMGs and snipers from the top of ANP Hill.

'The boss went into the operations room with the OC and emerged some ten minutes later with a new mission for us. With no down time and a very quick pat on the back to all my lads we "prepped our kit", changed radio batteries, rearmed, forced cold food down our throats and huddled round my map. Our new mission was to counter enemy movement to the east by setting up an ambush on a nearby wadi and road.'

Less than forty minutes after returning to the safety of the District Centre, Corporal Hewett was forming up his section ready to return to Now Zad's killing fields – but this time it was his turn to lay an ambush. They were still covered in Taliban excreta.

As he led his patrol through the narrow streets, mortars continued to target the enemy while the Apaches and Harriers had moved away to refuel, leaving one B-1B bomber watching overhead for enemy movement.

Approaching the ambush site, Hewett posted the left and right cut-off groups from the two other sections within the troop and, although they had specific and pre-ordered tasks, right now they were, primarily, an early warning for 1 Section and troop HQ with the Joint Terminal Attack Controller (JTAC) and the Mortar Fire Controller (MFC) – the 'killing group'. Rear and flanking protection was to be courtesy of fire from the District Centre's sangars – small defensive pits topped off by stone or sandbagged low parapets – and from those on ANP hill. The B-1B bomber remained overhead.

'As we set up the ambush I opened my air panel marker to ensure the B-1B had "eyes" on our position and sent grid references to the JTAC who passed them on to ANP Hill and the bomber. Meanwhile the boss sent our location and situation report to the company operations room within the DC – which they relayed back to the Joint Operations Centre (JOC) in Camp Bastion.'

Sadly this mission was something of an anti-climax, for the Taliban had clearly appreciated that they had failed to succeed against the British in a straight firefight, so they resorted to moving away, mingling with civilian groups, an action that prevented the Apaches or 'fast air' from engaging them. 'As the hours rolled by, nothing happened. The enemy were assessed to be moving well to our east along a much larger wadi; however, as we had no positive ID on the targets we could not allow the B-1B to engage. Quite often the Taliban would move around with civilians and children to protect themselves. So we waited until the operations room ordered me to collapse the ambush and move back to the compound. As we returned, a Chinook was inbound to resupply us with ammunition.'

The next day, unknown to Corporal Al Hewett, Marine Dan Ravenscroft wrote a one-and-a-half-page personal letter to his company commander, Neil Sutherland, which stated:

> I am writing to request that the actions of Corporal Hewett on the date of 8/12/06 be noticed by the higher ranks because of acts of extreme bravery, courage, unselfishness and calm leadership while under heavy fire from an enemy ambush…I hope that all can appreciate what he's done…
>
> Yours sincerely,
> Marine Ravenscroft.

Major Sutherland explains his decisions of that day. 'I did not have the combat power to take them on in this place and I had to bear in mind the considerable number of reinforcements coming in. I had no choice, but I had drawn them out and could now positively identify many of their positions, so that was a positive result for the patrol. The risks outweighed any potential gain. Had we sustained three or four casualties we would have had considerable difficulty getting them back across the open ground: we could not have brought a Chinook in as they really were not that keen on landing at "hot" landing sites. So I simply established a baseline and hit them hard with a couple of Apaches and some Harriers as we were moving back.

'What I didn't know until the debriefing was that Corporal Hewett's section had taken considerable fire from a couple of compounds. We went back to them a few days later as we had identified them as future targets. So the Taliban had broken the ceasefire in a big way as the terms allowed us to carry out these framework and familiarisation patrols, which is what we were doing when we were attacked.

'The elders were upset to a point. They tried to say that I had crossed the boundaries but I told them that no boundaries had been agreed: we were not targeting anything and once I managed to persuade them that it was clear who opened fire, that it was clear who broke the ceasefire and it was clear that it was not in our or their interest for me to have done so – they admitted that it was the Taliban.'

The marines believed that Now Zad's Taliban commander had come under external influence higher up his own chain of command.

'It may have taken a ceasefire of three months before the civilians would really start to come back but I was already achieving a certain amount of success and was hoping to continue to build on that, but now we were back at square one. Before that we had established, through the elders, a "cash for work" scheme whereby we paid them to clear the place up and to help us do the same: and we were actually achieving some success. Colonel Matt and I believed that our success was a significant factor – maybe not the only factor but a significant one – in the Taliban breaking the ceasefire. They actually saw us winning the civilians over to our way of thinking.'

The Taliban had broken the ceasefire, curtailing Kilo Company's essential work of reassurance. Neil Sutherland – and especially the men of 1 Section – knew where the firing positions were and with anti-ambush drills very much part of the company's training, for so often the only way to entice the Taliban into action was to 'trail one's coat', it was time to go after them.

The second patrol was planned for 12 December and was aimed

at the wadi to the east of the desert scrub, towards the two compounds from where Al Hewett's men had taken serious fire before. During the intervening four days Desert Hawk, the unmanned aerial vehicle launched by a huge rubber catapult, had shown that the wadi was permanently occupied. 'The wadi offered a relatively covered approach; we appreciated that it should have been mined although at the time, mines and IED – at our local level – were not used that much. A positive indicator in our favour was the regular use of the route by locals.'

The company, in an 'advance to contact' formation, moved as a slow caterpillar, north up the wadi – knowing only that they would be ambushed before the day was out. The Fire Support Group – FSG – were in the rear, ready to move to an appropriate flank once action was initiated. Company TAC HQ was in the middle behind the leading troop. Earlier, Harrier aircraft had over-flown the general area just to remind the Taliban what was in store should they come out into the open.

Any expectations that things would go the same way as on the previous patrol – although this time with more than just one section taking incoming and fighting back – were quickly dispelled. The Taliban were evidently prepared, for the company was ambushed from five separate positions simultaneously, with the enemy in significant numbers using small arms, RPGs and mortars. Major Sutherland recalls, 'It happened rather unexpectedly. The FSG, at the rear, was shot up quite badly with one casualty; Marine Richard Watson, the vehicle commander, was shot through his side, the bullet striking him beneath his combat armour. Marine Freer, the driver, grabbed him and held on to him as the Pinzgauer, with one shredded tyre, staggered to a halt. The FSG commander extracted the vehicles back down the wadi and out of the killing area while 5 Troop instantly coordinated 81mm mortar fire in retaliation from ANP hill plus smoke and HE.'

Marine Watson was carried back to Coy TAC but the nature of

his injuries categorised him as a T1, and he was, in Neil Sutherland's words, 'in a really, really bad way'. The commander was faced with a difficult decision; should he pull back, and evacuate Marine Watson, or press on and hit the targets from where they had been ambushed? 'Now I had a serious casualty on the ground with a bad military situation, as we were in danger of being surrounded. It was a case of "Do I go for one of the positions and we fight through – take more casualties – or do we extract back to the DC and live to fight another day and, quite frankly, give Marine Watson the best chance of survival?"' Unfortunately, as Sutherland says, he 'did not have enough combat power to prosecute any of these targets while at the same time carrying out a life-threatening casevac' – so in this instance he decided to pull back. In addition he knew that the company, which had been based at Now Zad for some time, were not moving away soon – 'We were going to be there for another two months.' Part of the reasoning behind the patrols was that they were fishing for information on enemy locations while not necessarily advancing to take up new defensive positions. Sutherland had known that they would not always be able to prosecute serious attacks because of the lack of combat power and thus he acknowledged the difficulty of safely extracting a wounded man. A few months later things were different; on *Operation Knight* at Kajaki the company suffered four casualties in one day and still carried on.

Marine Richard Watson was evacuated back to Camp Bastion where he was, sadly, pronounced dead. 'We kept him going but he probably died while the troop was picketing the wadi during the move back to the DC,' said Sutherland. 'We didn't always pull back,' he added. 'It was not a set policy to stop any advance just because of a casualty. It was always up to the commander on the ground and was a balance of many things, especially risk versus gain.'

Marine Watson's death signalled that the ceasefire was well and truly finished and left 42 Commando's marines with their first fatality. Despite this tragic setback, and in the face of increased

Taliban presence in the area, Sutherland had to continue negotiating with Now Zad's displaced inhabitants. 'After that we tried to maintain links with the elders as best we could but there was an increased reticence to come to the DC. A couple of times I met them by invitation out in the desert; arranged through a local, if I could find one. I said that if I saw anyone with a weapon in the local market I would assume he was a Taliban and that they should get that message across, especially as here the locals did not as a rule carry weapons as they do in Arabia.'

After the ceasefire's collapse it became a question of stalemate, but Kilo Company was not going to stop patrolling. Every time they deployed from the District Centre it was as a company minus – during *Herrick 5* each company had two troops – that pushed out as far as it could from the DC.

The company consisted of two troops and a Fire Support Group; the FSG carried HMG, GPMG SF, Javelin and GMG – 'which is a fantastic weapon, as is the Javelin'. The Javelin was popular not just because of its extraordinary range but also because of its incredible accuracy. The Taliban would hunker down inside their compounds, which – apart from the disintegrating one at Taunton – were usually so thick that they were about as good protection as they could get from small-arms fire. Javelin, though, could be put through a door or a window, which meant that the marines would know for certain that whatever they shot Javelin at, it actually hit; if they wanted to hit inside a compound, then they could.

Always popular, of course, were the Apache helicopters. Their range meant that they could loiter for a couple of hours overhead at Now Zad, depending on weapons or fuel loads, as well as the time of year; but the Taliban had – or were reported to be *about* to have – ZSU anti-aircraft weapons. One or two Apaches did take some rounds through their fuselages but, said Sutherland, 'When we were there none was shot down. The RAF were quite loath to supply us with Chinooks but we did get some vehicle resupply or

C130 airdrops of ammo and mission-critical kit – dropped the other side of ANP Hill over Christmas when we were very low on ammo. But occasionally the parachutes did not open ...'

In January Kilo Company was relieved by Lima Company, who had also driven up in Vikings: as one company moved in the other moved out, allowing Sutherland's marines to relax for two days at Camp Bastion while some even played a game or two of cricket.

Way to the south, in Garmsir, life was equally busy for the newly formed IX Group. Colonel Magowan always knew that the area might fall to the Taliban at any time, an event that would be judged – rightly – as a strategic failure, and especially so after the 'loss' of Musa Qal'eh, back in November. Since September 2006 Garmsir had nearly fallen three times to the Taliban, the brigade staff believed; that it never did fall was due to 'a splendid Royal Irish captain called Doug Beattie' working with the ANP and an ad hoc group of Afghan Security Forces plus a handful of UK forces. Between them – and thanks to superb leadership – this tiny force had managed, by the skin of its teeth and hand-to-hand fighting, to stop the Garmsir District Centre from capitulating. Beattie was to be awarded the MC, but not before the staff realised that they had to invest in the Garmsir area: if the DC was lost the coalition would no longer control the 'psychological gateway' to Lashkar Gah, the capital of Helmand.

There were other reasons for investing in Garmsir: it was the convergence of an estimated fifteen supply routes across the desert, up from the southern border with Pakistan; intelligence showed that along these nebulous tracks men and *materiel* were trafficked northwards while key players regularly travelled back and forth from Quetta. Garmsir was the natural place for newly arrived insurgents to join the descriptively named green zone and, thus, the more secluded and near-secure Taliban pipeline that led not only to the provincial capital, but also the economic centres – that is, the poppy collection and distribution points – and, obviously but importantly, military concentrations. The green zone is a fertile band running the full length of the Helmand, within which the Taliban considered, with good reason, they had full freedom of movement. It was close terrain, criss-crossed by narrow tracks, canals, irrigation and drainage ditches, all affording excellent cover from view and excellent cover from fire. 'Once in there it was very difficult to back out again.'

From the outset the brigadier had made it clear that in addition to the Afghan Development Zone he needed to 'do something about the desert', but he also knew that he did not have enough forces to hold ground or maintain a persistent footprint. He could, though, interdict and disrupt the Taliban's supply lines and freedom of movement using the Mobile Operations Groups formed around the Light Dragoons and the Brigade Reconnaissance, plus, occasionally, Zulu Company in the Vikings. Thomas's brigade needed to identify enemy target concentrations, especially those around permanent suspected Taliban bases such as Jugroom Fort and Koshtay.

As well as the security considerations, Garmsir was geographically important. A road runs into the town from the western desert, crosses a substantial bridge on the outskirts before bisecting the built-up area, west to east, then stops on the bank of the north–south canal to the east. It was this road, intelligence reports suggested, that formed the front line; anything to the north was government-influenced and 'reasonably friendly' to British forces, whereas all to the south – a myriad of compounds and dusty alleyways – was Taliban-controlled. To the east, government influence stopped at the canal; beyond this waterway was hostile territory. With the Taliban exercising freedom of movement up to the road, the sparsely defended and derelict Agricultural College, just to the north, received continual 'incomers'.

To accord with the brigadier's wishes, the Brigade Reconnaissance Force was despatched south. Captain Jason Milne and his Brigade Patrol Troop had had an interesting start to their tour. 'We deployed on the ground within four days of arriving in Helmand and took over the 16 AAB's Pathfinder – the equivalent of the BRF – vehicles, WMIKs and two Pinzgauers as HQ vehicles, but they were not in good nick – not the Paras' fault as they had had a pretty torrid time. We also inherited their kit but not their terms of reference. Our remit was to patrol from Garmsir, effectively the front line, south to the fishhook. Generally out in the desert to the east of the river; to the west the ground was easy going but there was nothing there. The east was harder terrain with lots more Taliban. Our predecessors had told us that it was not a good thing to get involved in the green zone or the towns, as there was always a danger of coming across the enemy, but our view was that that was precisely why we were there. But a good point is that Brigade Reconnaissance Force, by its nature, should not come into contact – we are a recce force, eyes and ears. But in the Garmsir area we were the first for some time although there had been some OMLT guys there.'

In the area they patrolled, the BRF quickly discovered that the desert had one major drawback for a troop conducting a reconnaissance patrol, even a mobile, armed one: 'You can't recce without coming under fire as there is always someone who will see you coming.' This led the BRF to adopt a new policy; 'We would drive forwards until we came in contact; probably return fire but our aim was to log and report what was going on.' Further north almost any kind of contact would result in air support of some sort being called in, but from Garmsir down to the fishhook 'no one would come, so our revised plan was to advance to contact, return fire, retire, think about it and return to sort the problem out'. This became their daily pattern.

'So we built up a picture for the brigade. Some of the contacts were quite heavy and would last a few hours and some were just a few rounds here and there. Although it might have appeared that we had the edge on the enemy because we were mobile, the Taliban, in the desert, did have an advantage. If you sit in a tight, well-defended box you can see someone coming from miles and miles away across the desert in all directions and that gave them time to prepare against a known enemy – us – from a known direction. Whereas we did not know what was about to happen.

'The Taliban would not leave the green zone once they got there. We assumed that all their supplies were coming across the desert, but that was not really proved. We had no means of covering the whole area down to the border as that was out of the range of helicopter support for ammo resupply and casevac. Actually we did go further south – just – but only after agreement with Brigade. As we came into contact every single day there was always a chance that one day we would need that helicopter. We took risks but Helmand is a risky place. Our remit from Brigade was to "find" and once we had "found" a host of missions would emerge such as: Interdict, Dislocate, Disrupt, Destroy or Neutralise.

'And you could only do any or one of those tasks once you had found the bad guys, and that wasn't difficult. As soon as we drove within Kalashnikov range – about 500 metres – of the green zone we came into contact and they would initiate a firefight. Then we would need to get closer to find out where they were and in what strength. Despite seventy-two serious engagements we had no casualties other than two guys shot in Garmsir as we were coming back from patrol, shot by the ANP who thought *we* were Taliban. We were on foot and it was dark. Sergeant Steve King was hit five times and an Irish marine, known as "Johnny Poo Tash" due to his inability to grow a moustache, twice. Steve woke up seven days later.'

The number of contacts the BRF received came as a surprise to Brigade, when Captain Milne reported back as instructed. 'I was told to report every single contact as Brigade wanted a better picture of what was going on, so I did it for forty-eight hours – only to receive an intelligence report saying that, "things have really got bad in Garmsir as the enemy have upped the ante with so many contacts". I sent back a signal saying that, actually, that is what it has been like over the last month but as you have now asked me to report every contact, I am doing so.

'To support the move into Garmsir itself and Zulu Company's MOGs, we needed a gun line in the desert; we recced a position and helped move the gunners into what they called FOB Dwyer, in the desert to the west and far across the bridge that leads into the town. The bridge was always a bit dodgy as it was the only way in and out; we were shot at every time we crossed, so I would drop off two WMIKs at one end of the bridge ready to fire and we would drive through them and the .5s would open up on the enemy positions. Once across, the first two WMIKs would fire the others across. And that became the standard drill.

'Most of the Taliban here were probably poppy growers, Tier 3, who were paid to take shots at the Coalition Forces; every so often you could see the difference when a professional was in charge or in the area. Then the mortar fire would be more accurate and we could get pinned down by these guys – the Tier 2s and 1s – which we weren't always. The Tier 3s were just sent to the edges to fire a few shots and disrupt. Our job was to go out, find them, as there was no one else to do so until the arrival of Whiskey or Zulu. Then we handed over a whole bunch of target packs showing that we had been contacted here, here and here and they took on what was in effect the front line in Garmsir; manned the checkpoints and improved them from being pretty primitive to something substantial.'

Before arriving in Garmsir to establish his IX Group's forward operating base – FOB Delhi, a year or so later the place where Prince Harry would conduct some of his forward air controlling when his regiment, the Life Guards, served in Afghanistan – Colonel Robert Magowan had asked, 'What are Zulu Company and C Squadron doing?' Chris Witts of the Vikings explains: 'We were operating as a Mobile Operations Group with Zulu Company, supported by C Squadron. But we were not just their mobility asset, we were their life-support system for food and water. Completely mobile and covering vast amounts of ground, we would recce villages and be a presence on the ground. The Brigade Reconnaissance Force was working further south and conducting recce by fire, getting involved in many firefights.'

Marine Adam Edwards was a Zulu Company radio operator with 1 Troop. Reckoning that a marine's greatest asset was his ability to 'yomp for miles into battle', Edwards had not been sure about the Vikings: 'Manoeuvre on our feet is what our training is all about and what we are best at.' His view quickly changed to one that considered this new addition to the corps's armoury 'a hoofing bit of kit'. Edwards had joined Zulu four months before the deployment, since when he had spent weeks 'MOGing' in the western desert with just a few contacts but no real 'battles'.

Chris Witts saw it differently: 'Each MOG would last for about three weeks and then it would be back to Bastion for four or five days before returning to the desert. If aircraft couldn't fly in the resupply I would send three vehicles back 150 kilometres to pick up the stuff. The beauty was that in the open desert we didn't follow tracks, so the chances of mines were lessened.

'We were MOGing in the south where we felt that we were having a substantial effect as we could pop up and disappear back into the desert to pop up somewhere else. We took our resupplies plus fuel and ammunition with large trucks so that we could stay in one area – totally self-sufficient and self-protected.

'We would establish a hub, which itself could move, and patrolled out of that, all round the western side of the Helmand River. We conducted one operation on the eastern side as there was a village we wanted to influence. This involved a drive for an hour and a half to find a crossing place, then, after we had done the job, back the same way. The intelligence guys usually told us if a certain place was good or bad, but they weren't sure about this one, so they asked us to have a look. Some villages we could approach on foot and some very definitely needed an armed recce, depending on the intelligence assessments.'

To launch his hoped-for domination of the green zone, Colonel Magowan decided that the Taliban had to be swept out, so he planned a clearance operation to the south of Garmsir. If he achieved his buffer the colonel could then begin interdicting down the Helmand River. 'Garmsir was the front line and the west–east road might as well have been the Somme or Ypres. Eventually we were to hold the checkpoints at JTAC Hill, Eastern checkpoint and Central checkpoint and we were to hold the DC and live in the old Agricultural College but south of that...if you went south of that road you were going to get shot at. This was the playground!'

JTAC Hill was to Garmsir what ANP Hill was to Now Zad: a mound on the western edge of the town by the river, it was permanently manned with a fire support team and when needed, and, as its name implies, a Joint Terminal Attack Controller. Its main task, as in the north, was to offer over-watch and flanking fire support for the District Centre.

Robert Magowan planned to use Zulu Company for his clearance operation on 5 December, with C Squadron in support. On foot, three troops would cross a start line to the west of the bridge and then move, in file, down the road until lined across the width of the town. Once in position they would turn right and clear for one kilometre to the south. When he was satisfied that the Taliban had been pushed south or killed, the colonel would then order the company commander to fall back to the Agricultural Centre. Here Magowan would assume responsibility from the OMLT, having been all but besieged there.

As with all operations, the colonel ensured everyone understood what he wanted from them by constructing realistic models in the dry desert earth based on photographs supplied by the RAF; in other circumstances, the men would be called round a table to see the layout of the ground they would be fighting across. Shallow trenches were dug to represent the river and its tributaries; pebbles and sticks stood for the buildings, with blue counters to show the compound walls while white and orange tape marked out the start line and the objectives it was expected the marines would reach. Once the colonel had finished his briefing, the section leaders and others who needed to know the exact positions they would be attacking and manoeuvring around studied the model and worked out what they needed to pass down to their own men.

With Zulu Company based in the desert, about seven or eight kilometres to the south-west of Garmsir, it was Chris Witts's task to move the attacking force company to the start line. A mixture of compounds, run-down buildings and piles of rubble, sand and rocks lined the road from which they would move south. When satisfied that his three troops were stationed down the road, evenly spread between the bridge and three checkpoints – JTAC Hill in the west by the river, Centre checkpoint and Eastern checkpoint by the canal – the company commander would order them to begin the advance in mutually supporting bounds. One troop would move forward and take up firm positions, supporting the next troop who did the same, and then the next – thus 'pepper-potting' troop by troop in a fire-and-movement manoeuvre. As a plan, it was well worked out and thorough; however, there were no reserve forces.

This was to be Magowan's battle group's first real test of combat command, and Zulu Company's first real test of close combat: it would turn out to be 'a twelve-hour epic with some unbelievable bravery from individual marines fighting through built-up areas and compounds'.

After a desert reveille of 0230, Zulu Company loaded into their Vikings and set off for the Line of Departure at the western edge of the Garmsir bridge. Only structured to take twenty-four tons at a time, the Vikings would wait until the marines were on the other side before themselves crossing individually and not in convoy.

Three would then line the road through Garmsir as a quick reaction force while another three secured ground to the north-west.

As the early mist swirled up from the river banks, and a wan light came through the dust, the order to advance south was passed down the chain of command. The ground the men began to cross was a mixture of compounds and cultivated fields – some vegetable, some poppy – with irrigation ditches criss-crossing them. Tunnels and holes linked one compound with the next, a maze of rat-runs. It was ideal ground for defending; a nightmare to attack.

'The operation started well with the Light Dragoons down the western bank of the river,' Chris Witts observed, 'plus a section of Vikings forward down the eastern flank. We eventually had three sections of empty Vikings behind the company – just in case.'

Picking their way through the rubble-strewn ground, the men were cautious. The light grew stronger and the early haze burned off; there didn't seem to be any sign of resistance yet. As 4 Troop overlapped 1 Troop and moved close to the first compound walls, on target for the next stage of the operation, the Taliban reacted with small-arms fire. The marines responded by forcing the enemy back, across and through the low buildings, allowing themselves to reach their first objective, report line Arbroath. Having secured that, they moved towards their next, an irrigation ditch named – for the purposes of this operation – Taunton.

The marines encountered, among the buildings, opium factories, but could do nothing about them other than check they weren't harbouring Taliban, for they had no personnel to deal with them. The level of fire they faced intensified as the Royal Marines closed in on Taunton, the Taliban rained down mortars, set off RPGs and poured small-arms fire on the approaching commandos.

Chris Witts, sitting back from the forward line, saw the strength of the enemy's fire; but then he and his Vikings were also targeted. 'I was at the Eastern checkpoint with Corporal Bob Northcliffe's section when a sniper engaged my vehicle with tracer only a few feet above our heads. As we couldn't locate the enemy, this sporadic fire continued for a few hours, and intensified when we took some round strikes to the front of the vehicle and were engaged with GPMG and RPGs. One mortar fell in between me and the next-door Viking and I looked across at the driver. We both gave a thumbs-up sign as it didn't go off. They were never very accurate and the first rounds were usually miles out unless they were lucky.'

1 Troop, with Marine Adam Edwards, was closest to the river for the 'advance to contact'. As the company reached the southern part of Garmsir town, the men had fanned out, with 4 Troop in the middle and 5 Troop furthest to the east. The left flank closing in on Taunton, they had continued to pepper-pot forward in turns, successfully avoiding the Taliban's sporadic fire – but as the enemy fell back through their own, familiar, well-established defensive positions, the rate of fire suddenly became 'horrific'.

The advantage the British had over their enemy was, of course, their freedom in the air. With Apaches and Harriers overhead, and the Light Dragoons watching their thermal imaging screens to the west, the marines could be warned of enemy positions as well as likely targets ahead.

'Just short of Taunton, the "maverick call signs" – that's what we called C Squadron – radioed us to warn of enemy movement. So they put down fire just in front of our position to help us move forward while our Joint Terminal Attack Controller, Captain Al Cairns, who was with us and not on JTAC Hill, controlled the Apaches.'

1 Troop had now reached report line Taunton's ditch where Captain Cairns turned to Marine Edwards and said, 'The Apaches are running out of fuel and have to go. I think this is where things might get interesting.'

He was right; the attack helicopters left at what was probably the moment they would be needed most. Within moments of the Apaches departing, their distinctive noise fading away, the Taliban launched their offensive. The parapet of the ditch that sheltered 1 Troop shattered along its length, bursting into fragmented stones and sand as a hail of small-arms fire came cracking in and RPG rounds thudded home. The occasional 107mm shell crashed nearby, throwing up dust and clods of soil that came clattering down on the sheltering marines.

Corporal 'Scouse' Mullin's section, on the far right flank, was manoeuvring through a crop of maize that ran along the river's edge. The cobs, brown and rotting, had not been harvested and, although well past their prime, the high stalks and wilting leaves still offered protection from view if not from fire. Seeing the difficulty that his neighbouring section was now in – unable to move from their ditch until someone could suppress the incoming fire – Mullin brought his GPMGs and 84mm unguided anti-armour rocket launcher, the ILAW, into action. A Royal Artillery bombardier, also stuck in the ditch with Adam Edwards, began calling in artillery from the gun line, ten kilometres out in the desert, with the three-gun battery dropping 105mm shells 300 metres to Zulu's front. It became Edwards's job to differentiate, for his company commander, exactly what it was that was exploding around and in front of his men, such was the rate of both enemy and artillery fire. It was not always obvious.

The Taliban kept on pounding the trench and Cairns barked instructions into his radio, before turning to Adam Edwards. He said, 'Tell your OC that I've got two F-18s on their way in from a carrier. Better tell him to pass to all the lads that this is going to be a "danger-close" mission.'

Edwards, crouching down, crept over to Captain Will Clark, the troop OC and Sergeant David 'Taff' Morris, to pass on what the JTAC had just told him. Obviously the marines would be too close for the F-18s to use anything other than their 20mm M61 Gatling guns. Edwards went back to his position next to Captain Cairns and answered a call on his radio; the men on JTAC Hill asked if the marines were now firing back behind them, as they were taking incomers. He barely had time to realise that this meant there were Taliban to their rear when Captain Cairns shouted that the planes were on their final run in. Edwards clamped his hands over his ears, waiting for the blast.

The planes shot past so fast it was almost all over before anyone had a chance to realise they were upon them. First there was the impact of their passing, a wind so strong that the men were thrown left and right by it; even the biggest in the troop, the GPMG gunners, were knocked backwards. At the same time the fighters' cannon were strafing the targeted positions, but the sad fact was that the ditch in which the marines were lying was too 'danger-close' and, as the 20mm rounds pounded into the ground around them, tearing up the dirt and flinging debris everywhere, shrapnel from one round hit an unintended target – one of the marines. Finally, as the planes soared away into the distance, there came a roar, a louder noise than anything they'd heard before.

Adam Edwards had been lying next to Captain Cairns when he felt a hammer blow on his arm, that flung him round so violently that he found himself facing the other way, staring at one of the corporals in the trench. He looked quickly at the top of his left arm, to see it hanging loose with blood pouring from a wound between the shoulder and the elbow. Shock set in quickly as the lads gathered round, applying field dressings; there was a dazed air about the whole place, as if everyone – Taliban included – needed to recover from the impact of the F-18's attack. The wound wasn't as bad as he first feared; the round had passed through the flesh of Edwards's arm without striking the bone. Someone took his radio and Adam Edwards found himself propped up, awaiting casevac, feeling, in his words, 'rather redundant'.

Except that Edwards had been shot by the Taliban, not by 'friendly fire'. Along the trench from him was Marine Jon Wigley, who had been hit in the back by one of the 20mm rounds. Those who had not gathered round Edwards to help him were quick to attend Wigley, but it was too late; he had been killed instantly.

For the members of 1 Section not immediately involved with Jonathan Wigley and Adam Edwards the battle was growing in intensity, with the Taliban now so close that they had begun to hurl petrol bombs. Scouse Mullin's section in particular was a target as the enemy tried to burn them out of the maize field, forcing them into the open. There was the added confusion of the contacts coming in from the rear, as the Taliban tried to encircle them.

The priority for the wounded was getting a casevac helicopter in nearby and organizing a Viking to come through the battle and collect the men. The call went out over the radio but the initial report of a T1, which meant the casualty must be seen within the hour, as he cannot look after himself, was changed to a T4, which meant either walking wounded or dead; in other words, there was no rush any more and no point in risking further life.

The Vikings did arrive, and put themselves between the firefight and the troops, one lifting the wounded back to the casualty aid post by the bridge, from where they were transferred into a Light Dragoon CVRT. This then took them across the bridge and on to the helicopter landing site, a safe distance into the desert.

'Before boarding the Chinook,' said Edwards, 'I asked the doctor if Jon was dead and he just said, "Your friend is not very well," but once we were in the air they pulled the blanket over his head – and that was a bit final.' Back at Camp Bastion Marine Edwards was treated by the nurses, for whom he had nothing but praise, and eventually recovered from his wound.

The decision was taken to pull the rest of the men back. Some didn't agree; the ICOM chatter they picked up suggested that the Taliban regarded the Vikings as some sort of 'tank' and felt they would not win against the British when they had this vehicle on the battlefield. But Zulu Company was stretched in two ways: they were lined out across the town from the river in the west to the canal in the east, while at the same time the company was seriously under-strength. With the capped manpower imposed by the government reduced further by injuries and R-and-R, 5 Troop on that day, for instance, was just sixteen men and not the thirty or so it should have been.

The Vikings drove out of contact, again keeping themselves between the small-arms fire and the marines, who jogged back alongside them. Elements of 3 Troop were also pulled back this way, as they were being shot up to such an extent that, as Chris Witts put it, 'they were crawling back on their hands and knees'. Eventually everyone was recovered.

Colonel Magowan's men had exploited as far south as he had planned, they had taken a friendly-fire casualty yet, despite the ferocity of the Taliban reaction, only Marine Edwards had been shot by the enemy. His Zulu Company had stirred a hornets' nest which was not to be resolved that day, so IX Group's forces were withdrawn to the Agricultural College, where 'We were mortared, shot at and shelled all night, which didn't do the locals much good as they were on our side and it wouldn't have been happening if we were not there. So the villagers fled north or just somehow buggered off out of the way.'

Witts's Vikings and Warrack's Scimitars joined them inside the college. 'We couldn't move in there with a company of marines, thirteen Vikings and sixteen Scimitars. Sentries were posted on all the corner points while we hastily built sandbag defences.'

So the IX Group spent its first night of many in the Garmsir District Centre, after a day that should, perhaps, have gone rather better. Command of the company was not improved by the time of the Jugroom Fort raid in January but with a subsequent change of command, Zulu was to achieve more convincing successes.

Company Sergeant Major Steven 'Shep' Shepherd had nothing but praise for his men, and didn't blame the American pilots for the loss of Jon Wrigley either. 'I never felt angry about the pilots. If we had not had them there that day we would definitely have lost more than one person. When things are going wrong it is fucking scary and the young nineteen-to-twenty-year-old marine is not looking over his shoulder at the young nineteen-to-twenty-year-old troop commander and saying, "What the fuck am I going to do?" He looks to the thirty-year-old "stripey" – troop sergeant – and they were all magnificent. Actually all the young second lieutenants were bloody brilliant as well but they didn't, yet, have the experience. I should know as I trained most of them at Lympstone! If there is a well-planned and well-led attack, the guys will go and keep on going with as much violence as possible – that is what they are trained to do and it is superb to watch. At one stage the Taliban were just twenty metres away and were even throwing petrol bombs at the lads but Sergeant "Spinner" Spence was as brave as fuck, as was Corporal "Scouse" Mullin, who held off a serious Taliban assault – and so on.

'I had great respect for the colonel, as he said before the attack, "Your job is to close with the enemy," and I like that attitude, as that is what wins battles. It is sad if by doing so you lose people but that is what command is about and something you have to live with and be unemotional about it, not for ever but during a battle. I felt guilty when "Wiggers" died, and sad, but I thought, "Fucking hell, we are in the middle of a scrap here, the whole company is in the middle of a scrap and although he's dead there is nothing I can do, so let's concentrate on fighting and winning." I felt guilty later because for the two hours after he was killed I never had a single thought about "Wiggers" – then, when I did, I thought, "You callous bastard".'

Now that his IX Group was firmly established in Garmsir's less-than-salubrious Agricultural College, Rob Magowan was unequivocal about his position and future. 'We were then shelled both directly and indirectly on a daily basis while JTAC Hill was being shot at all the time by 107 rockets and RPG. So what was I doing? I was going to protect the Garmsir District Centre. I was going to hold that come what may, for the influence that the Taliban would gain if they took the DC would be significant.'

India Company, described by the brigade's chief of staff as a composite company from 45 Commando Reconnaissance Troop, Air Defence Troop and other command support group ranks, formed the Lashkar Gah Operations Company and, as a back-up force, were in many respects always the bridesmaids and never the bride. On permanent call to help out wherever it was deemed necessary, they were summoned to Garmsir by Colonel Magowan in order to relieve the more formally constituted Zulu Company for larger operations. Nevertheless, so efficient was India Company that to enjoy the 'Lashkar Gah effect' became a synonym for receiving expert and timely support.

As has been mentioned, the bridge to the west of Garmsir was the only route in and out of the town and so was prone to sniper fire whenever it was crossed. Captain Tom Evans-Jones's men more often than not found themselves protecting the bridge. While some contacts were minor, most were drawn out and – often – fierce. On one occasion, after the marines had dropped a salvo of 51mm mortars and called in the Royal Artillery's 105mm guns, the Taliban managed to sneak up to positions beside the bridge, getting as close as 200 metres. As the defensives here were only shallow trenches with a foot-high sandbag wall, it wasn't surprising that some of their fire was impressively accurate.

Marine Eddie Cain had narrowly missed being shot here only three days before. The precision of the Taliban fire forced him back, then, peering over the defensive wall, he felt something flash past him – very close. 'Rolling back from the trench I realised that a round had gone through my shirt, just missing my shoulder. The second time in three days.' Spurred on by irritation, perhaps, Marine Cain – under the cover of returned fire – raced down to the vehicles, which were also under attack. 'Jumping into our WMIK I shunted back another vehicle, so that I could bring the .5-calibre machine gun to bear.' The WMIK's .5-calibre had a great range and real punch. 'All the time more RPGs and small arms were coming back at us but as soon as there was a lull I fired into the Taliban's compound. That stopped them.'

Captain Tom Evans-Jones was 'particularly impressed by the control the marines displayed; they were able to switch in moments from aggressive actions to winning the hearts and minds of the locals, which is essential to what we were trying to achieve. We had to provide stability so that the local population could rebuild their lives in this once thriving town without fear of the Taliban.'

On 10 December, a sad ceremony took place beside the airstrip at Bastion. The commandos, standing in rows three deep, in their green berets, gathered to send Marine Wigley home. A stiff afternoon breeze blew as his coffin was carried aboard the plane for his journey. In the January 2007 issue of the *Globe and Laurel*, and at the end of a more formal obituary, Marine Laws of Zulu Company was to end his tribute with the words, 'Marine Jonathan Wigley, you were a hoofing bloke and a hoofing Bootneck. Words cannot describe how much everyone will miss you. Goodbye and God bless.'

Despite all his units being fully occupied throughout Helmand during December, Brigadier Thomas was required to mount a brigade operation – *Operation Baaz Tsuka* – in Kandahar Province by providing supporting elements for a Regional Command (South) task. If Thomas deployed an important force into the Maywand District this would entail considerable risk for Helmand; nevertheless, *Baaz Tsuka* was an important operation, for the loss of Kandahar could presage the loss of 'the south'. Consequently the brigade was ordered to act as a deception followed by a disrupt-and-block phase. A secondary aim, understood by the brigade staff, was to show a demonstrable commitment to the wider coalition.

Various participants offer their own views on 'combat camping', as the marines were to call *Exercise Baaz Tsuka*. Colonel Holmes explains, '*Operation Baaz Tsuka* was a Regional Command South deployment which the brigadier wanted to support as a task force operation. We deployed from Camp Bastion with Juliet and Lima Companies and my TAC HQ plus the Estonians – all of us in either Vikings or the Estonian vehicles and established a Forward Operating Base eighty kilometres away. My mission was to interdict the enemy and prove to him that we were the main effort – the main threat to him – coming in from west to east.

'Meanwhile the Canadians and other task forces were in the vicinity of Kandahar ready to strike. We went there expecting a heavy fight but the Taliban had evaporated. I thought we would be employed on compound hits but the intelligence was not very good as the targets we were actually given were nothing as described. We certainly disrupted them but it was not as expected. We conducted the operation between 14 and 24 December, established an FOB and took a very large amount of combat power and combat support. We established a large footprint with company group patrols spreading out, hoping for the opportunity to engage as often as possible with the locals. This was one of the successful aspects as we gained an immense amount of situation awareness which we could pass up to regional command HQ.

'We changed our patrols to engage with the elders and the local civilians so we could garner support for ISAF and the Afghan government. This included distributing to the villages essential materials such as wood for buildings, shovels and tools as well as, inevitably, footballs. Our families in the UK had sent out hundreds of pairs of shoes for us to give to the children. It was heartening to see the remarkable flexibility of my men, who one moment had been spoiling for a fight but the next were fitting sandals to small feet.'

From his perspective, Major Ben Warrack of C Squadron reckoned that, '*Baaz Tsuka*, in support of Task Force Kandahar, is not worth a serious mention. It was a bit of a damp squib although to begin with we were all shitting ourselves because all the intelligence suggested that we were going to be severely taken on by the Taliban – and we weren't. We came back on New Year's Eve down Highway One, which I always avoided, except on this occasion. Much better for us to go across the desert, but the decision was taken at about ten o'clock on Christmas Eve in the evening that all of us – the whole brigade battle group – would get on the road and drive back to Camp Bastion. There must have been over two hundred vehicles, maybe three or even four hundred. We were at the very back and it was dark and took for ever. Before the start we were warned that we would be mined and ambushed. They mine from the side with a sort of Claymore mine so the fact that it was a hard-top road didn't matter – we were just as vulnerable. And actually there were large sections of the road that were not tarmacadammed and of course there are numerous culverts – just like Northern Ireland. And because it paid off no one ever talked about it again.'

Major Steve Liddle, commanding Lima Company, felt much the same. 'In the middle of December we were told that we were being sent to an area full of hard-line Taliban whom we were to "destroy", not simply "defeat". We were warned that we would be fighting our way through enemy positions and at least five compounds. It was the anniversary of Limbang when Lima Company under the command of Captain, later Major General, Sir Jeremy Moore led a successful assault in Borneo to rescue British hostages on 12 December 1962, and so I did my bit and said "This is exactly how the guys would have felt before Limbang," but the reality was different. Ewen Murchison and I conducted a recce in a Black Hawk and flew over the whole of our target area, but nothing looked hard-line to us; unconcerned women and children with no defences. On our return we told the CO that there was something wrong. The information we were working on had us kicking down doors when we should have been giving them footballs.

'We patrolled for two weeks and only had one RPG fired at us, otherwise nothing, not even during the epic convoy back. On our return to Camp Bastion on Christmas Day there were loads of parcels from the Royal Marines Association and the housewives of Devon. The lads wanted for nothing – apart from booze and girls. Two cans of beer and we all felt the effects of that.'

Despite frustration, in many quarters, at being dragged away from what most considered to be their real area of operations, *Baaz Tsuka* was a notable success, for it was important that the Helmand Task Force was seen to play its full part as a coalition player across the whole area. Coalition feathers – which could be, and often were, easily ruffled – had been smoothed, allowing the commando brigade a little more freedom of movement in the next-door province.

December in Helmand was also a busy time for Mike Company. With the Taliban threat in Kajaki increasing, Colonel Matt Holmes believed that one under-strength troop was unlikely to contain the security situation, especially as that troop was unable to patrol. Its duty was to man the observation posts at Normandy and Athens but it had hardly enough men to do even that. So, in accordance with the brigade commander's wishes, Holmes surged Mike Company Group into Kajaki in early December.

Outposts Athens and Normandy were occupied, as was the 'ex-United Nations base' at the bottom of the hill by the river. This now-dilapidated compound was below the dam and close to the late King of Afghanistan's summer retreat; this had once been rather more opulent, with stone pillars and a swimming pool, but it was now occupied by an OMLT team of two 'green beret' SNCOs from 29 Commando Light Regiment, Royal Artillery. The FOB itself, now named Zeebrugge, lay approximately 800 metres west of the Kajaki Dam and was overlooked to the south and east by mountain ranges, two of whose peaks were occupied by friendly forces and one by the Taliban. These peaks dominated the area. Fifty metres to the north of the buildings, the Helmand River ran westwards and could be crossed only by a locally built bridge, just downstream of the dam. To the north of the river, the country was undulating and dotted with small villages, towns and numerous wadi systems. The dam was therefore surrounded on three sides by a strong Taliban force, while to the rear of FOB Zeebrugge lay the huge and vital reservoir.

On the night of their arrival, Mike Company's marines found that it was normal for FOB Zeebrugge to experience nightly engagements with a regular exchange of mortar fire and heavy machine guns. Throughout December, Mike Company would be occupied making their locations safer with, at last, some overhead protection from the nightly assault of RPGs, 107mm rockets and small-arms fire. Collin's men also pushed out to the north, across the bridge and into Tangye which, although unoccupied, would often produce a surprised local – but no Taliban.

The power station, downstream from the dam, is at the end of a road that follows the river's southern bank in from the west, past an American compound where two civilians managed the operation of the hydroelectric plant; then on past the District Centre, the jailer's quarters and the jail itself. Across the river and north of Tangye the police station was surrounded by empty buildings, which Mike Company also cleared as a matter of precaution. One of the main aims of Mike Company during its posting there was to create a buffer zone between the dam and the Taliban, so that the civilian workers based at the dam could carry out their jobs in relative safety.

The nearest locals were at Chinah, some four kilometres north-west of the dam and on the edge of a wide, north–south wadi. Those living to the east of Chinah found the Taliban 'moving among them like locusts', taking their money, their food and often forcing the elder sons to fight for their cause. As Tony Forshaw explained, 'Once they had sucked a village dry they would move on to another one. They were charging the villagers in Kajaki Oliya electricity money – though there was no charge for the electricity anyway.' In practice the dam was working but only to a fraction of its full capacity which, if achieved, would supply power for over two million people across the province.

Mike Company first took over the two high points that commanded the area, using the heavy weaponry at their disposal; this allowed them accurately to pinpoint the Taliban positions and, thanks to the increased ranges they now had, to fix the Taliban and suppress them. Next the company carried out a series of patrols that allowed it to dominate the ground; along with 59 Independent Commando Squadron, Royal Engineers, they built a permanent vehicle checkpoint at a crucial road junction that enabled the local Afghan National Police to police the area better.

'With that physical footprint established, the company was in a good position to begin a series of deliberate attacks and company-level fighting patrols aimed at forcing the Taliban away from the dam,' said Captain William Mackenzie-Green, the officer commanding 10 Troop. 'But the enemy are doughty fighters: despite suffering heavy casualties they almost always chose to fight on rather than cede ground.'

Some weeks would pass before the marines felt that the skirmishes they were involved in had worn down the Taliban sufficiently for the buffer zone, between the enemy and the dam, to be truly effective. Mike Company would be conducting recce patrols for the whole of December while gradually pushing the Taliban further and further out, and, as their footprint expanded, they would come into regular contact yet were always able to rely on fire support from Unknown Left, Normandy and Athens. 'Fast air' and Apache were also called in, especially after a contact from a mountain top to the north of the river nicknamed – for good reason – the Shrine. Two 1,000-pound bombs brought an end to that until it featured again during a deliberate attack over the New Year period. By the end of December Martin Collin and his men had built up a good picture of the Taliban positions and were shortly to be in a position to strike with confidence and considerable aggression.

Meanwhile Whiskey Company remained in FOB Price until Christmas, from where they conducted patrols in Gereshk, back-filling for Juliet Company who were conducting an MOG further up the valley in their Vikings. Danish CIMIC teams, working in the town, had identified certain potential building sites for schools and hospitals. When they had time, the marines of Whiskey Company joined the Danes on the ground, allowing them to talk to local leaders in order to enable building projects to start. Gereshk was beginning to enjoy relative peace: a previously unlikely hope, for it is at the southern end of the lower Sangin Valley – real 'bandit territory'.

With others returning from *Baaz Tsuka*, Whiskey Company could, on Boxing Day, redeploy back to FOB Robinson, four kilometres south of Sangin town, where they were to remain until the end of the tour some three and a half months later. Their task was to reassure the local population, protect the outpost and the two 105mm light guns of 29 Commando Regiment that supported the District Centre in Sangin. They also interdicted Taliban supply routes and movement corridors that ran through the area. FOB Robinson, named after a Canadian soldier who had been killed there, was not a popular station, for it was on the edge of the green zone with a large, loose outer perimeter and a tighter close perimeter.

The brigade commander was keen that it should be developed as a patrol base and Colonel Holmes could see why, although still regarding it as a poor location, for it is surrounded by dead ground that allowed the Taliban to come in close and lay mines. Marine 'Dutchy' Holland of Whiskey Company was killed there by an explosive device. The company also had a Land Rover which went straight over a landmine within 200 metres of the 'gate'. The engine block was blown over the top of the vehicle, but luckily the two marines escaped with only light injuries. On a daily basis, 102mm mortar and 107mm Chinese-made rockets were sent hurtling in by the Taliban, but serious damage was limited thanks to the large number of sandbags placed around the base. Their chief characteristic was the noise that preceded their arrival; a shrieking sound that meant, to British ears, 'take cover'. So for a while there was a ban on whistling, in case anyone became confused.

Conditions inside Robinson were equally unsatisfactory. Accommodation was a difficult choice between a retired P&O shipping container or one of a few mud huts, each of which needed considerable and continuous cleaning even to approach a basic level of hygiene. Washing was possible only through a combination of bottled water and a pump rigged to a well. Electricity was provided, sporadically, by an ancient Canadian generator 'discovered lurking in a corner' while nutrition was courtesy of the interminable ration pack 'A' – hamburgers, beans, chicken pasta and treacle pudding.

On 17 January the first company patrol left under Captain Rob Money's command with a mission to conduct an area of operations familiarisation while interacting with locals in order to pick up any information, or 'atmospherics'. As they moved through the outskirts of Sangin, the ICOM chatter increased and the pattern of life began to change. Men, women and children were running away while vehicles, on spotting the patrol, were turning around and driving off at speed. Money made an instant and obvious decision, ordering the company to leave the area immediately. The company turned south, heading towards a different section of its operating area.

Suddenly, from only thirty metres away, the Taliban opened fire. Three RPG rounds were lobbed at the Pinzgauers but, amazingly, missed all of them, while small arms tore into the ground and pinged off the metal fittings. Inevitably, at this point the signals failed, and so only sporadic communication was possible between the WMIK fire support group and the Pinzgauers who were trying to reconnect with the rest of the patrol. Meanwhile two Apaches circled overhead, shooting where they could but staying clear when it was too confusing for them to get a clear view of their target. They were, though, able to observe the Taliban massing around the Pinzgauer group, up to 150 men pinning them down.

Now the Taliban had to contend not only with the Apaches in the air but the rest of the patrol in their WMIKs. The fire support group launched their counter-attack with relish, and although the Land Rovers had on many other occasions proved not to be the safest place during a firefight, their heavy weapons gave them the edge that day. Marine Ian Danby, perched behind his heavy machine gun, didn't stop firing, even when bullets penetrated his vehicle. Finally, the 105mm guns in FOB Robinson launched 160 rounds, 'danger-close', which helped keep the Taliban at bay.

Pretty soon the Apaches ran out of ammunition, but by then two F-18s were on their way, called up by Captain Tom Vincent and Bombardier Myers from 29 Commando Regiment, Royal Artillery. Their arrival, after the patrol had been under fire for over an hour and a half, bought time for the men to withdraw. Except that as they did so one of the Pinzgauers stalled. CSM Robert Jones ran back to the machine and, despite being shot, managed to get it going again. The Pinzgauers and Coy TAC were eventually able to rejoin their comrades.

Marine Ian Danby won the Military Cross, CSM Jones was Mentioned in Dispatches, while Captain Money worried that he had jeopardised the Sangin 'ceasefire' by being too aggressive.

5
JANUARY: DYNAMIC UNPREDICTABILITY

*'Right at the top there was a bunker. We could hear a commu-
nications radio inside, so the room was cleared with a grenade.
As the radio was still working when we entered, we gave it to
our interpreter and from then on we could hear everything they
said. One group was discussing our presence. "The Brits are
coming ..." "How many?" "... I think they have sent all of them
for us ..."*

Mike Company's sojourn at Forward Operating Base Zeebrugge
fell into two phases. For the first month the company acquired the
all-important 'situational awareness' with the surrounding area and
the local friendly forces, while the two hilltop OPs – Athens and
Normandy – began to identify enemy dispositions as well as many
frequently used firing points. Most of the enemy locations were
revealed at night when sentry posts, military training areas and
movement with weapons were observed through the company's
extensive range of night-surveillance equipment: the enemy having
failed to appreciate the power of the company's night-vision capability. A series of night patrols determined whether the enemy
possessed any surveillance capability at all by contrasting the
Taliban's reaction to the night patrols with those conducted during
the day, made by comparing SIGINT activity as well as enemy

contacts themselves. Despite a twelve-day operation elsewhere that restricted patrolling for a large part of the period, by the end of December recce patrols were being deployed for close observation of the areas of interest identified by OPs. They were also able to collate a mass of intelligence from which to plan offensive operations: approaches, potential fire support positions and so on.

Sergeant Jay Layton commanded Mike Company's Fire Support Troop on Normandy, 'where we took over from Kilo Company and had brilliant observation to the north over all the villages. The guys from Kilo had, understandably, been slightly fixated by Sparrowhawk, which wasn't yet in our hands. When I took my team in we decided to change the men around regularly so they could get a feel for the whole area and not become fixed to one spot. After about a week in the three small buildings – bunkers really – that had been built on top of old Russian defences, we had begun to make ourselves a little more comfortable – and safe. We placed barbed wire around them plus Claymore mines and had a very good idea of what was going on in the local area: for instance we identified the Taliban training camp in a town to the north-west of our position. We watched the guys coming in at night and practising fire and manoeuvre and bringing in heavy equipment; we also identified four blocking, or sentry, positions in various towns. We studied their routine carefully, then passed our findings back to company HQ and Major Collin. He would make his assessments and although we could have engaged them all it was decided to continue building up the intelligence picture before going for a surgical strike. We wanted to take them out *in toto*.'

To confront the daily contacts they were receiving, the company commander began sending out patrols from FOB Zebrugge; happily these contacts were limited because the OPs were at the limit of the effective range of the RPGs, often lasting no more than fifteen minutes; sometimes, however, the contacts were drawn out over four hours. Mostly they were regarded as a nuisance rather than anything more dangerous; not only did the base have top cover by then, but because they were beyond useful range for the RPGs it meant that the grenades exploded harmlessly in the air, almost like fireworks.

The marines operated the Leopold telescopic sight system, which allowed them not only to observe the Taliban but also to have immediate read-outs of trajectory and distance to the target. Desert Hawk was used to look more closely at key ground features, enemy locations and approaches. They still had, of course, access to the enemy's communications systems, so would listen in when they observed them setting up weapons such as their 107mm rockets on Kalawak Ridge. Doing so enabled them to understand the code words the Taliban used for their various systems – classified but given such names as Ladyfingers and Watermelons. More usefully, the bases and OPs that the marines occupied were more heavily armed than before. If the Taliban tried to approach the marines, either on foot or on motorbike, when they were about 1,500 metres away they would be hit with everything the British had – a wall of fire. Along with the .5-calibre machine guns, they now had grenade machine guns, which can fire 340 40mm rounds per minute up to 2,000 metres, and Javelin at their disposal. Plus, of course, substantial air support.

One of their tasks was to watch for the places that the Taliban used as bolt holes and firing positions; for instance the enemy would use small holes in compound walls to shoot through, but then disappear out of sight, so making it tricky to pin down accurately where to return fire. This was tedious, delicate work as the men had to squat or lie patiently, peering through 'eyepieces' till something caught their eye. One day, up on Normandy and clutching a cup of tea, Sergeant Jay Layton was watching through the 'scope when a whirl of dust attracted his attention. Next the glint of metal and the sound of a belt-fed machine gun confirmed it: some Taliban were lining up to take a shot at them. He called out the reference point and everyone in the OP reached for a weapon and started firing at the place where the gunmen were hiding, keeping them pinned down. 'There was a gap between the wall and another wall, which was their escape route. The compound is on a hill so we could look down the length of the gap between the walls, to carry on pinning them down.'

Sergeant Major Mick Smith, 29 Commando Regiment's FST commander, coordinated an air strike, calling in a B-1B bomber. The men on Normandy stopped firing, and waited to see if the bomb flushed out the enemy; only if they showed their faces again would the men shoot. The plane's bomb landed at the top right of the compound, completely destroying that. For a moment, nothing happened, then three men in their black robes scuttled out of their hiding place towards a building higher up the slope. They'd chosen badly, however, because the men in the OP could see clearly all the exit points of their new hide so, while they waited for the B-1B to make its return pass, they just observed.

'The B-1B came over again but the pilot said he couldn't, for some reason, just drop one bomb; he had to drop two together and we said, "Yeah, no problems!" And they were straight on target.'

During this first phase, significant effort was invested in building a good working relationship with the local Afghan friendly forces. There was no civilian population as such for, thanks to Taliban coercion, all locals had evacuated the area rather than face being caught in the crossfire between Taliban and the British. Friendly forces, beside the British, were disparate and included the District Commissioner, the Afghan National Army, the Afghan National Police, the Helmand Security Force and two American coordinators with their own Afghan security force. An OMLT team with ten ANA was already in place prior to Mike Company's arrival. Coordination was vital so that maximum use could be made of all Coalition Forces while ensuring that patrolling could continue without concerns for 'friendly fire'. Enemy contacts around the town of Tangye were a daily occurrence with mortar and 107mm rocket attacks on the garrison increasing.

The second phase of the deployment was marked by the transition to offensive operations and a more robust domination of the area of operations. This had been planned from the outset and, following a surge operation, was mounted in the New Year starting with the occupation of the third mountain-top OP, Sparrowhawk, on 1 January; an assault and clearance of a hill known as the 'Caves and Shrine' feature on 2 January; the installation of a permanent vehicle checkpoint on 3 January; and a fighting patrol to an identified Taliban training camp on 4 January. The intention was to show the Taliban that the Royal Marines had the power to hit them hard, and to demonstrate – not only to them but also to any civilians who might be in the area – that the marines meant business, and would not rest before completing their task. Each of these engagements was fierce and prolonged, and it was thanks to the coordination with the other branches of the British forces in Helmand that they achieved all they did. The value of employing Apache attack helicopters as flank protection and to ambush withdrawing enemy forces – the leakers – was to be confirmed time and again.

The main aim of *Operation Clay* was to provide stability in the area, so that the dam could be renovated without the civilian personnel working on the project coming under attack from the Taliban, as they had realised what a threat its success posed to their position in the region. 'The whole business was not just about building or replacing the second turbine – due to arrive from China – for there is a NATO plan to build an airstrip and a road,' said Colonel Matt Holmes. 'In many respects, the whole "Kajaki Project" epitomises the very reason why we are in Afghanistan.'

On 1 January the first part of the plan was enacted, as the company secured Sparrowhawk, a dominating piece of ground with 'fantastic arcs south, west and north'. Athens and Normandy had long been occupied by the British but Sparrowhawk, the westernmost of the OP positions overlooking the Helmand River below the dam, had remained a thorn in their side as it had, in turn, been occupied by both the Taliban and the militia. Athens and Normandy had come under regular fire from Sparrowhawk, so it was considered a vital piece of ground to hold as well as protecting the two eastern OPs.

The Recce Troop and the Royal Engineers went ahead, clearing or marking 'legacy' mines as they went, not just to ensure a safe passage for the men who were to occupy it but also to 'influence' the few locals who were around by demonstrating that the British were there to help them. Another troop had been sent to the northern part of the Helmand River to prevent any possibility of the Taliban staging a counter-attack. On top of Sparrowhawk, old Soviet Army defensive positions were taken over by the men and improved. Not surprisingly, the Taliban were unhappy and fired rockets and RPGs on to the OP from the village of Kajaki Olya, to the south, where they enjoyed freedom of movement. They also attacked Mike Company's other troops, including men deployed to the intended permanent vehicle checkpoint, down the hill to the west of Sparrowhawk, who became engaged in close-quarter fighting among the low mud buildings either side of the road. The attacks on the team that occupied the OP continued well into the evening.

Martin Collin had been given a 'long list of tasks', including dominating the heights, influencing the locals, improving the capability of the ANA and the ANP in the area – making sure they even spoke to each other – defeating the Taliban and enhancing force protection. 'The number of small operations we conducted considerably reduced the number of attacks on our positions on the ridge line from Kajaki to the westward, based on the principle that attack is the best form of defence,' Holmes said.

With Sparrowhawk denied to the enemy, the accuracy of some of the Taliban attacks was reduced as they no longer had the advantage of a good arc of view, nor the use of the site as a firing point. At last the Royal Marines could launch a series of operations to the north of the river now that there was greater over-watch and added flanking fire support. First on the list for Martin Collin was the clearance of the 'Caves and Shrine' feature two kilometres to the north-west of Sparrowhawk, across the southerly bend in the river and above and beyond a wide, north–south wadi nicknamed the M1. At the hill's western end an ancient shrine could still be discerned, while caves riddled the eastern end. The position was used, regularly and persistently, as a firing point by the Taliban, despite being engaged with mortars and air-dropped munitions every time fire erupted from it. The mission was straightforward: 'Take the Shrine, destroy any enemy that resist and show the Taliban that there is no hiding place from the commando brigade.'

The night before the assault, 11 Troop, who were to be the point troop, lay up in a derelict police building where they prepared for their pre-dawn attack. 11 Troop's commander was Second Lieutenant Bertie Kerr, who, only a month earlier, had completed training before coming straight out to a front-line appointment. His troop sergeant, Pete McGinely, supervised the building of a communal fire in the only enclosed room with a roof – of sorts – then persuaded his marines to heat their 'boil-in-the-bag' bacon and beans in a gigantic kettle found lying in a corner. Once fed, Kerr encouraged his twenty or so men to get some rest, as they were going to be up at 0400. It was difficult to sleep, though, when the noises outside included the occasional burst of small-arms fire. The thoughtful mood was broken when one of the marines, Thomas Curry, known as 'Vinders', announced, 'It's my twenty-first birthday today.' Congratulations were offered – one marine passed a hip flask of single malt – and Bertie Kerr, at twenty-three all of two years older, added, 'Congratulations, Vinders – and a Happy New Year as well.' Curry thanked him, then said, wistfully, 'I really wish I was at home with the girlfriend right now, going to the pub.' The silence that followed suggested that he was not alone.

11 Troop's mission was to draw fire as a moving, advancing trip-wire which, once sprung, would give the security forces the targets they needed for their mortars and Apache gunships. By and large tripwires are expendable but that was not how Bertie Kerr had briefed his men: 'We advance until spotted and if that is close it will be straight into the assault, but if we still have some way to go we move to a flank – I'll decide at the time which one – with 10 Troop giving covering fire from the valley behind us. Then the mortars and choppers will do their bit before we fix bayonets.'

At 0445, in the early, bitter mountain air, and in darkness, the company crossed the departure line before making its way slowly up the hillside, heading towards the top of the slope at the south-ern end of the target area. The men kept in a line, remaining apart so as to prevent getting easily shot at, until Bertie Kerr signalled to them to stop – 'Go firm.' They gathered together and sank down to the ground, keeping a low profile in the gathering light, hoping that they wouldn't be spotted before they could see their objective – but then *crack, crack* – and the green glow of tracer bullets came flying up towards them.

They'd been identified by Taliban in the compounds in Shomali Gulbah, a village one kilometre beyond the Shrine, and those first few tentative shots suddenly turned into a neon blizzard as the pre-dawn sky lit up with tracers. With the bullets creeping closer to their targets, the enemy unleashed RPGs and mortars. The marines, though, had accomplished part of their objective by drawing out the Taliban so now the WMIK-mounted FSG could hammer into the enemy with their heavy machine guns. 11 Troop manoeuvred toward the exposed ground to the north-east of the objective where once again they were pinned down with little cover. The mortars behind them were ordered to lay down smoke, and with a great, dense cloud hugging the ground ahead the troop could move off. Before they did so, Kerr issued the order to fix bayonets, but the satis-fying sound of the blades clicking into place was followed by a hideous *snap* and a muffled cry; Marine Richard Mayson rolled on to his side, clutching his left arm. Shot in the wrist, he was jabbed with morphine, escorted down the hill and eventually to the field hospital at Camp Bastion.

Father Michael Sharkey, 42 Commando's much revered and admired Roman Catholic padre, always tried to make it his business to be where the bullets were thickest, for he believed, understandably, that that was where his ministry would be most appreciated. That morning, well before dawn, he had joined the company sergeant major, Wayne 'Taff' John, and two medics to follow the company's advance. 'We parked up by a wall and all of a sudden I saw my first RPG about ten feet above me and then masses of small-arms fire. I looked up and the whole dawn sky was lit up with tracer. At that moment we received a call that one of the men had gone down: Marine Mayson. When we reached him he had been bandaged up by his mates and he asked me to sit and talk to him while we waited for transport. We had just had the Christmas carol service a few days before at Kajaki and as his sister and mother where in the choir at home he started to think of them. I could see then that he was "going down", not because of the injury to his wrist but just the fact that he had been shot. He asked me to hold his arm up and to cup his fingers. Which I did and slowly got him comfortable until the Pinzgauer arrived. He was full of morphine by this stage as well. Back at the Kajaki "sickbay" he was so cold that we put him into a sleeping-bag but that didn't really do the trick so I had to get in it with him!'

Eventually a Chinook arrived and he was lifted to Camp Bastion hospital with the padre still accompanying him. On the way, Richard Mayson asked Father Sharkey if he would tell his parents before they heard any rumours. The Padre tried to convince him that it was he that should talk to them first as he was not in a life-threatening state, although the wrist bones had been seriously shattered.

Michel Sharkey explained his reason: 'I told Marine Mayson that it would be best if his parents heard the news from him, otherwise they might be a bit frightened if the first voice they heard was that of 42 Commando's padre. But no, he insisted. So I phoned his home and explained very quickly who I was. Told them that their son had been involved in an incident and that he had been shot in the wrist and was going to make a full recovery. There was a long silence and then his father said, "Thank you so much for letting us know – so much better than someone coming to the door in uniform." Then there was another long silence until his father asked which wrist it was. I said it was his left wrist, to which his father replied, "That's all right then. Thank goodness it wasn't his drinking hand." I knew then that all would be fine with both parents and son.'

Dealing with Marine Mayson's injury at the Shrine hadn't slowed up either the attack or the response; as the marines in the valley provided covering fire, 11 Troop crept closer to their target. About 1000 an Apache arrived to hurl missiles into the midst of the Taliban positions. 11 Troop charged forward with bayonets fixed to find many Taliban dead, and signs of others having fled.

The Shrine was cleared of all trace of the enemy, including any weapons and ammunition, before the troops linked up to withdraw across the exposed M1 wadi under cover of smoke. However, instead of returning to FOB Zeebrugge, the men of Mike Company moved to a new Line of Departure ready to assault and clear the village of Khalawak. It was from this northern flank area that small-arms fire had been received during the assault on the Shrine. It was also known to have been abandoned by ordinary civilians to the Taliban and now needed sorting out while there was still momentum to the day. This eventuality had been anticipated, so an Engineer section had been attached to each troop to ensure that they could gain access to the compounds through using 'mouse-hole charges' and bolt croppers.

The impact of the bombs falling around the marines created a haze of dust that rose off the walls and the rubble, almost like steam rising around their feet. The sections pressed up against the compound wall and took up their positions; two kept watch, checking along the wall in case any enemy suddenly appeared around a corner or over the top. The din had changed now – no longer did the bullets whiz past but instead the *crack-crack* of small-arms fire was stronger – and the regular thudding of mortar rounds was reflected back off the walls, swamping the men with the noise. It was useless to shout orders and instructions at each other, so hand signals were used.

With the mouse-hole charge placed against the wall, the sections pulled back from the site, keeping their weapons up and making sure that each member of the team was covered. As soon as the charge had detonated in a hail of stones, brick and dirt showered them all. One man peered carefully round the hole, weapon raised, before slipping through, sliding over the rubble on the other side to be in a safe position – his back to the inside wall – to provide cover for his fellow marines. He was followed by the GPMG gunner, who would ensure heavy fire if the marines needed to protect themselves against an aggressive Taliban counter-attack.

Entering any kind of structure where they could at least see ahead was bad enough; going into a building where they couldn't see at all was more dangerous. The men didn't know if there was going to be someone in there, maybe wounded, waiting to die – and wanting to take some of the enemy with him as he did so. They'd line up outside, then the point man would ready a grenade, reach quickly round the open doorway, and pitch it in. The explosion would send shards of rock flying out through the open doorway, followed by a cloud of black dust. Two men would watch the back of the building, waiting to see if anyone might try to escape. Almost immediately the first man was running in, weapon up and ready for whatever he might find inside, one man from the section sticking like glue to him while the others kept watch outside. If the room was particularly big – or if it was dark, thanks to the dust everywhere – they might pop off a few rounds first, just to be on the safe side.

The push through the compounds was exhausting, because it was a very tense business, waiting outside a wall, then charging in once the opening was made, never sure what was going to be waiting on the other side. But the men dealt with each contact well and no more injuries were sustained. At the end of a very long day Martin Collin's pleasure was radioed around the company, 'Well done, guys. You can count today as a success and on the Taliban's home ground as well. You have shown them that we will take them on wherever we want to and not be beaten back. But there is still much to do if we are to achieve our aim of returning the area to peaceful prosperity.'

The next day, with the Shrine no longer a threat to them, the company commander turned his attention to the permanent vehicle checkpoint that had originally been intended for installation at the same time as the capture of Sparrowhawk. However, as not all of the construction stores had arrived, it had to take place sequentially. The need was threefold: it protected the western approach to the dam and the FOB along the only access road and thus vehicle-borne weapons getting close to Sparrowhawk. Sparrowhawk, in turn, provided high-ground cover for the PCVP itself. The overt military presence at the checkpoint would also act as an interface with the local population while extending the security footprint. Throughout the night the Royal Engineers worked hard to complete the task within a robust security perimeter supplied by Mike Company. The third of January had been a less busy day when compared with the first two of the month, yet there were still lengthy engagements for the screening forces to the north and for the Sparrowhawk OP.

The following day, 4 January, the PVCP was still under construction, so, while providing area protection for that work, Martin Collin took the opportunity to exploit the momentum. Just over a kilometre further south down the road they were working on was the village of Kajaki Olyao, close to which, it was assessed, was a Taliban training camp. A large fighting patrol was dispatched. The men were to gather in a group of trees that lay across the fields from the compounds that made up the camp, after making their way there in the usual open lines. As they were about to move off towards their start line, machine-gun fire from the compounds drilled into the ground around them. The men dropped back towards the relative safety of the trees while the officer commanding made some hasty calls to Sparrowhawk; as a result of which mortars and heavy machine-gun fire from the OP sprayed across the camp, allowing the marines to press on. The first assault section dashed across the open ground with a barrage of fire support to suppress the enemy's fire, and were quickly followed by the next section; together they pushed into the first compound, with the enemy falling back as they did so. The close-quarter fighting was intense; the Taliban could not use their RPGs at this range, so they were mostly reduced to small-arms fire. They would appear to be shot, only then to drop out of sight behind the wall they were firing from, so forcing the marines to lob high explosive or phosphorous grenades over the walls and into the buildings, to clear the way ahead before a bayonets-drawn assault.

At this point the remainder of the company came under heavy fire from their southern flank: a thunderous noise of RPGs, landing in and among 11 Troop's gun group. 10 Troop whirled around to suppress the fire coming from the south but they too were drawn into fighting at close quarters among the compounds. The remainder of the buildings were cleared and searched, before the company could withdraw back to the FOB.

Mike Company's scraps with the enemy were not over but, further south, a raid of a different, more audacious nature was about to be launched. On 10 January the Brigade Patrol Troop, a sub-unit of the Brigade Reconnaissance Force, under Captain Jason Milne and supported by Major Ben Warrack's C Squadron Scimitars, was tasked with a covert operation that they unofficially code-named *Operation Talisker*. Milne, who hails from the Isle of Skye – 'I like the whisky of that name' – is a mountain leader and had joined the corps in 1988 as a marine. Commissioned from the rank of sergeant in 2004, he regarded commanding a troop of the Brigade Reconnaissance Force as the 'best job in the Royal Marines'.

The objective was a suspected Taliban command-and-control node at Koshtay, a large village way to the south and on the east bank, just north of the fishhook where the Helmand River takes a suddenly westerly turn. Positive voice identification of a Taliban command centre had been received, so Colonel Magowan was confident that this was a genuine target, with ten buildings in Koshtay highlighted as a Taliban base; the main compound of which had once belonged to the area governor. Closer inspection by covert observers, hidden along the west bank of the river, made it clear this was indeed a Taliban HQ, for they had watched vehicles moving in and out and black-clad men on mopeds coming and going. Experience told them it was also a viable target.

IX Group's CO asked Milne how he suggested 'sorting this out' as the compound was on the east side of the river. From the western desert, he would have to cross the river, but if he approached from the east he would have to negotiate the whole width of the green zone.

'The river was the Taliban's western hard shoulder and they relied upon not being attacked from that direction, as it was supposedly impassable. We came up with a plan that involved small inflatable boats being sent down from Lashkar Gah that we would paddle across. However, it proved too difficult to get the boats to us so as we war-gamed it we realised that if we got stuck we would have to swim anyway. We did find a vehicle crossing point but managed to drown a Pinzgauer trying to get across the first time. Once we had recovered it we went back to the drawing board.

'I spoke to the CO and said I would like to try it again but this time wade across the Helmand, enter the green zone, cross a canal and then observe the target. If satisfied that it was a Taliban HQ I would bring in air and bomb the place before getting out the way we came. We thought that the Taliban would not expect anyone to cross the Helmand at night at that time of the year as it was cold, the river was 110 metres wide and quite fast-flowing – in fact it was in spate. I am reasonably small, so to me it was going to be chest-height but for some of the guys it was only waist-height. We wouldn't need to swim but there was always that risk.' A month or two earlier the men had watched civilians walking across at a particular point, when it wasn't in spate, so Milne knew it was possible. In fact, there was a form of submerged causeway, almost a ford, which the civilians had been using; but it was going to be near-impossible to find the exact spot at night and cross the river not just stealthily but covertly.

The plan was extraordinary. While tactical river crossings in arctic conditions are part of 'the elixir of life to a mountain leader', it was very, *very* cold – the air temperature at night was -6°C. However, Milne, having been granted clearance to conduct a covert recce, despatched a team to see if it was feasible. They managed to reach two-thirds of the way across but stopped for they did not want to alert the Taliban to the fact that it was possible. The plan was approved and Milne's BPT set off from the gun line in three Pinzgauers, accompanied by C Squadron's Scimitars. From the chosen lying-up point a patrol scouted ahead, then reported that all was clear to cross.

They stopped about three kilometres up-river on the west bank, before walking down parallel with the river until opposite the north end of Koshtay. A group of men stayed with the vehicles to provide covering fire during the crossing, should it be needed. The plan was for the twenty-four men to tie themselves together in four teams of six, then wade over. Normally any such crossing conducted at those temperatures would be conducted in wetsuits or at the very least, full Gortex suits; once they reached the other side, then they would pull on dry combat clothing before carrying on to their objective.

Not this time. As this was to be a raid, and nothing else, Milne decided that it was best to travel light to the crossing point where they would strip off, stuff their clothing, weapons and ammunition in dry bags, swim or wade in underwear and flip flops, then dress on the far side. They would continue the yomp to the target, call in the air strikes, and be away before the Taliban knew what was happening.

'It was quite amusing to look through the night-vision goggles to see a team of guys in underpants and body armour wading through the water; actually one of the guys had a pair of his girl-friend's knickers on which really did make a bizarre sight. He said he had brought them out just in case – but there you go – typical Bootneck,' laughed Milne.

The noise of the rushing river covered the sound of their wading, and by 0100 all were across. They dressed quickly, then moved off, cold and wet; Milne was thankful for their winter training as he felt certain that this had toughened up the men enough to be able to make such a journey. There were no signs of sentries on the river bank, but then why would the Taliban anticipate a strike at night from across a freezing and flooded river? Anyway, it seemed that the enemy didn't have access to the same quality of night-vision equipment as the marines, so they probably wouldn't have seen them till too late anyway.

The marines needed to creep close enough to the target to correctly identify it as a 'C2' node. On the ground they had nothing to rely on other than themselves and their own abilities, so they were out of reach of any ground recovery; but in the air, hanging back at present, they had Apaches, a B-1B bomber and a Nimrod on call.

It took them nearly two hours to cover the three kilometres

required to close in on the target area. Milne sent snipers forwards to take out any sentries, helped by the Nimrod above identifying the 'hot spots' through thermal imaging. This in itself was encouraging, as it meant that the attacking troop was receiving a live feed from the plane, so they were very much in control, even though they were deep in enemy-held territory. However when Milne spoke to the spotters in the aircraft, they responded, 'When will you be arriving?' Puzzled, Milne replied that they were already there. 'Negative, we have no sign of you on the thermal cameras.' Milne read out the grid reference, sure there was a mistake somewhere – and after a moment's pause the message came back: 'OK, we think we see you, but you're not showing up on the thermal cameras clearly, you're still blue. Are you still cold?' The marines hadn't warmed up enough from their freezing swim to register on the sensitive equipment high above.

The sentries, duly identified, were despatched at about 0400, before the men moved as quietly as they could into the rear of a compound, about 260 metres away from the target. This was the moment to observe their objective, although the sentries had in effect given it away. Once that was confirmed Milne hunkered down with the JTAC, Sergeant Jay Reed, to call the B-1B bomber. The pilot was ready to release the bombs; he only needed to know what grid references to programme into their guidance system and the quantity the target required.

The B-1B bomber came back and asked for a shopping list: 'What sort of bombs would you like?'

'What sort have you got?'

'Well – you can have four of these and two of these…'

The shopping list agreed, Milne explained that he wanted them all dropped in one pass as they didn't want to face the prospect of several dozen angry Taliban surging out of damaged buildings, looking for someone to shoot. It had to be conducted in a complete way: a 100 per cent kill first time. Once the bombs had done their work, the plan called for the twenty-four men to rush in and gather up everything useful – papers, photographs or anything of that sort – to take back with them; they hoped that any survivors left would be in shock after the weight of bombs that was about to fall on them.

As the aircraft banked for its dropping run in, the men on the ground moved closer, readying themselves for a charge into the ruins. 'We edged to about 200 metres, which was not good and certainly against all the regulations, as we needed to be 700 metres from a 2,000-pounder and I had asked for 12,000 pounds of ordnance.'

The bombs were released. Thousands of feet above the Taliban camp the B-1B pilot started counting down the seconds to death – ten, nine, eight … which Captain Milne relayed to his men, in a stage whisper via his headset. At the count of 'five', Milne said, 'Right, get your heads down lads,' but of course, being 'typical Bootnecks', once the count was down to 'two, one', all the men had their heads up to watch the devastation unfold.

The 12,000 pounds of bombs were all GPS-guided, so were superbly accurate, and exploded in one long, rippling *cruuuuu-ump*, that blasted through the compound, sending debris hurtling hundreds of feet into the air. The ferocity of the explosion shocked the men for a moment or two, a few whispered, 'Fuck me', as the heat wave smashed outwards, burning them at that short distance. The noise was simply stunning, the destruction almost total.

The attack was being recorded not only by the Nimrod far above but also by the low-flying Apaches, which swept over the area. The initial pass made by the helicopters was to clarify the situation on the ground before the marines moved in while Milne kept in constant contact to learn what they could see. He was also conscious of the time, for it was now well after 0400 and they still had to travel back the three kilometres to the river before swimming over to safety.

Those Taliban that had survived the blast – and they weren't many – were staggering around, in a daze. Flames crackled around them, as dogs barked furiously at the sky above . The Apaches returned to fire their cannon and Hellfire at any structures remaining, ensuring there was nothing left of the camp for the Taliban to reuse. When they ran out of rockets, they resorted to flechette. Milne decided it wasn't worth entering the camp now; the blast had been so devastating that very little remained and what did remain was burning. Rather than cross a small bridge over the canal to enter the danger zone, he decided to re-cross the river.

'The Apaches kept themselves busy for a while; there was plenty of work for them and they had a field day.'

The return to the river was easier as the men had no need to travel quite so quietly – after the blast they'd set off, there was little point – so they were back on the bank by 0500. Just before they reached the river they bumped into a civilian – clearly so by the colour of his clothing – in a state of some shock. Milne called the interpreter and told him to tell the man who they were and what they'd just done. He wanted to emphasise how easy it had been for the marines to get close to the camp, and to target it; that the Taliban's back garden was not a secure zone for him. 'And that we would be back. I thought it was a good opportunity for some PR regardless of whether he was with us, against us or neutral.'

Once more the men stripped off their clothes – giving one marine the chance 'to don his girlfriend's knickers again' – and swam back across the river. Waiting there were the men of C Squadron, ready to provide covering fire if necessary but also to give a running commentary on how it had looked from their vantage point.

Spurred on perhaps by the encounter with the civilian on the enemy bank, Milne took the opportunity to patrol through a local village. He wanted the villagers to be surprised to see men emerging from the swollen river as if it were nothing, having carried out such a destructive operation on the eastern bank. 'We wanted them to know that we were not to be put off attacking the Taliban under any circumstances. Now the Taliban knew that even the Helmand River was not an obstacle to us – something they had always thought it was.'

Once they'd returned to base, all Milne's men could share in the elation of a successful operation. It couldn't have been carried out so well any other way – the men had had to be on the ground to positively identify a Taliban target. Using all aspects of their training – the fitness levels demanded, the ability to withstand the cold temperatures – it was, in Jason Milne's words, 'a classic Royal Marine operation'.

If January was a busy month for the commando brigade, then 10 January was its busiest day. In Gereshk, early that same morning, Ewen Murchison's Juliet Company was embarking on what the marines were to call 'the day of days' at Habibollah Kalay, a clash that came to be known as the Battle of the Sluice Gate. Habibollah Kalay lies four or so kilometres to the north-east of Gereshk and had long been suspected of harbouring a major Taliban Command and Control Headquarters. Nearby was a sluice gate, where water was channelled from the Helmand River into the irrigation canal that runs to the south of the town; strategically important, it supplied the power via the hydroelectric station that had been the scene of heavy fighting during the *Operation Slate* series earlier.

Major Murchison's plan was to drive through Habibollah Kalay, the men in Vikings with WMIKs up front; 'bold, brazen' was the aim. Once through the town they would head to the sluice gate, establish a cordon, and then conduct a search for the suspected Taliban HQ. As usual with any of the patrols the Royal Marines conducted during their tour in Helmand, all the time they'd be waiting to see what reaction their presence in the area brought, what one marine succinctly put as 'tipping up in the back yard of the enemy and waiting for a reaction'. Ewen Murchison thought it was a 'little more sophisticated' than that, as the intelligence they had on the use of the building by the Taliban was good.

The patrol formed up at FOB Price, preparing themselves for the drive towards the town. Corporal Michael Cowe, who was later to win the Military Cross for saving lives after a Viking hit an anti-tank mine while in contact, described the formation. 'Juliet Company Group, including two 105mm guns from E Troop, 79 Battery, was split into a northern and southern group; the northern group consisting of 1 Troop, 2 Troop and Company TAC HQ, mounted in Vikings and WMIKs, while the southern group would establish a gun line to the south-west of Gereshk. 3 Troop – the Fire Support group – were tasked with providing over-watch of the target area and preventing reinforcements coming from Zumbelay about six kilometres east across the main river. Mortars would be based on the familiar spot heights of 836 and 852 – across the river and to the east of Gereshk – and at the same time able to observe across a wide area.'

This was, at the time, not only the largest operation that the company had carried out since arriving in the province, but also the first time a deliberate operation had been undertaken in the known hostile area of the green belt. Ewen Murchison was particularly keen, therefore, that the men conduct themselves in the most professional manner – not only to protect their own lives and those of their comrades, but also to ensure that the locals in the area saw a vast difference between them and the Taliban. Everyone prepared themselves for this difficult task; the troop sergeants checked, communications, ammunition and weapons. Unfortunately, Sergeant Willy Whitefield, 1 Troop, lost the bar armour protecting his side of his Viking cab, but worse met them outside the base when a body was found, blocking the company's path. A civilian contractor, reported missing, was lying in the road with three bullets in the back of his head and a note from the Taliban.

To Major Murchison it was important that he treated the incident with care: 'I had to deal with that properly, helped by the ANA who were with me, as they were to be our rear security for the operation.'

As the light grew brighter with the approach of dawn, 3 Troop, under Second Lieutenant Duncan Law, sat in over-watch on the spot heights. Corporal Russ Coles, whose WMIK would provide arcs of fire over the dam to the north, watched the night fade: 'The sun rose over the sleepy compounds, then signs of life began to appear; as dawn broke a layer of mist blanketed the ground, while the northern group continued its move along the canal to the target area.' Corporal Bailey's WMIK faced east, while further east still, on a spur overlooking the river, another FSG section was 'observing'. Also in the vicinity was Murchison's mortar line and the company second-in-command with the main HQ.

The job of the WMIKs in the convoy was to provide a cordon around the Vikings, ready to use their heavy weapons to provide support once the marines were out on the ground. One WMIK group of four vehicles was commanded by John Thompson, who was to win the Conspicuous Gallantry Cross to add to his Mention in Despatches from Iraq. He always referred to his 'waggon' as being 'like the United Colours of Benetton', as his usual gunner, Steve Davis, was half-Spanish – 'semi-gringo lad, top gunner, brilliant guy' – while his driver, Nathan Beagles, was a 'black guy from the south of London and a bit of a geezer'. In addition there was a heavy-weapons specialist assigned to him, as not only did they have the GPMGs, the .5-calibre, and the Javelins, but also the new grenade machine guns. Thompson himself had helped with the Arctic trials of this weapon, so was familiar with it. On this 'day of days', the heavy weapons would, at times, be used against targets at a 'nerve-wracking' ten metres.

Once the dead contractor's body had been carefully removed, the drive to the town of Habibollah Kalay went smoothly. The men were already able to see some signs of the impact they had made during their tour; earlier on, if they saw women and children, they would expect them either to run away – a sure sign Taliban were in the area – or have to fire a few warning shots to keep them back, as they didn't want to risk being among civilians if the enemy launched an ambush. Now, though, the locals seemed pleased that the Taliban's freedom of movement had been restricted, for the children waved at the men as they drove past.

The drive into the outer areas of the town halted at 0647 as Juliet Company, with Corporal John 'Tommo' Thompson's WMIK leading, came under fire at close range from the fields and scattered compounds either side the road. The ambush was massive, with a huge weight of small-arms, RPG and mortar fire from three sides – some as close as twenty metres away with black-clad enemy high up in the trees. Although not entirely unexpected – it was going to happen some time – the company was now pinned down along a single track. Rounds and bombs tore into 1 Troop's position while the marines immediately launched themselves into their well-practised anti-ambush drills. Corporal 'Tug' Wilson, a section commander, leapt bare-headed from his Viking into a ditch. Just as he did so an RPG exploded – harmlessly, thanks to the vehicle's armour – on the Viking. He shouted at the Viking's commander: 'Chuck out my helmet,' as it had been knocked off in his haste to climb out. 'Next time shut the bleeding door properly,' was the only response, as the Viking closed up.

For the next twenty-five minutes, the area was a solid wall of fire, light and noise, as the orange flashes of tracer rounds zipped around and over the men. Mortar bombs fell among them with a *crump*, while artillery shells thudded into the walls and the ground, smashing rocks and dirt and anything else they hit. Everyone was firing back, shell cases were flying; the armoured Land Rovers and Vikings took a severe battering – but held. The noise was a barrage itself, so intense and so constant that John Thompson burst an eardrum and was partially deafened. 'The noise of battle is the one thing everyone will remember,' Corporal Cowe said. 'You cannot hear anything, especially on the net, but something inside your body kicks in and you get on with the job in hand; everyone switches to autopilot and the training takes over.'

The marines of 2 Troop had also been attacked by RPG rounds on the canal, but they'd successfully fought back and either killed or driven away their attackers and were now seizing the opportunity to try and link up with 1 Troop by getting round to their right flank. This manoeuvre forced the enemy to turn and face them, thus easing the pressure on their colleagues, but 200 metres on they were engaged from two more enemy positions. The Taliban were mounting a major counter-offensive against the British; whenever a target was downed, another appeared to take his place.

The men in over-watch with Second Lieutenant Law were desperate to join the battle, yet as they took up positions to fire down on the compounds below, an RPG whistled a few feet over Law's Viking. Quickly tracking its path, 'ten to fifteen enemy 200 metres to the east reinforcing a well-defended trench' were reported; almost immediately a 'hellish rate of fire' was let loose by both sides, with enemy rounds striking the steering block and gun mount of Corporal Coles's WMIK, while RPGs screamed past Corporal Bailey's head as he shouted target indications to his gunner.

The trench was well located and the efforts of the .5-calibre gunner to kill the enemy proved fruitless. So Marines Talbot and Moncaster crawled out from behind a Viking, carrying a Javelin missile. They were quickly spotted. Soon rounds were pinging in the ground and whizzing in the air past them, but they stood their ground and locked on to the target – fired – the Javelin whooshed out of its tube and crashed home into the trench. Thanks to their coolness there was a brief respite in the contact but further along the canal the fighting remained intense. Corporal Mick Cowe returned to the job he'd started, looking down into the compounds and irrigation ditches below, and gave what he himself described as 'an impeccable fire control order' to Lance Corporal Coe. 'Bang the ILAW down there, John,' he said, and the rocket smashed into a wall, destroying one of the firing points.

Among the low buildings and fields, Viking and WMIK crews were providing devastating covering fire, being resupplied by other troops as they used ammunition at a rapid rate. Corporal 'Tommo' Thompson's WMIK, with Marine Beagles firing the Minimi and Marine 'Afghan' Steve Davies on the .5-calibre, were down to 20 per cent ammo; Corporal 'Adz' Lison on his GPMG was shouting, 'I've done five grand already.' An RPG exploded next to Corporal 'Kibbler' Matthews's section, shrapnel pitting his arm – at FOB Price it would 'require Savlon and a plaster'. Marine 'Fatboy' Farr, fully recovered from being trapped beneath an overturned Pinzgauer during *Operation Slate*, carried up extra link, muttering, 'We're fucked if we're pinned down here much longer; that canal looks fucking cold.'

After nearly half an hour of 'hoofing carnage', the shooting became more sporadic and spreading out over a wider area. The distance created allowed the company commander, his Viking also hit by an RPG, to 'bring in death' from above using mortars, artillery and air. 'All of these assets throughout were excellent, with artillery fire at one point hitting targets between the two Close Combat Troops. Well done, guns!'

The resupplying of ammo was being organised by the new company sergeant major, Marty Pelling, who had arrived in Helmand only six days before. 'My first trip was turning into the biggest contact the company had had to date, so that was a real baptism of fire for me. Fourteen thousand seven hundred 7.62 rounds needed to be resupplied in the first four hours, which was on a scale never seen before. I had 1 Troop's sergeant, Willy, and two of his marines coming back for ammo using the drainage ditches, then crawling back along them struggling with ammo boxes – known as liners – just as though we were training.'

By now it was daylight with attack helicopters circling the area and dozens upon dozens of empty ammo liners littering the area. Trees, bare in the cold of winter, had been cut down by the weight of fire, while around them compounds smouldered. 1 Troop, after fighting their way out of the initial contact point, were receiving Quick Battle Orders from Second Lieutenant Hughes to assault an occupied enemy compound. Corporal 'Tug' Wilson thought 'he might, at last, need his helmet'.

Under a heavy weight of suppressing fire, 1 Troop could now advance to conduct a 'hard knock' clearance of a number of compounds with high explosive and phosphorus grenades laid down ahead of them. This was dangerous and tense work; each enclosed building could harbour a ferocious enemy, happy to die for his cause if he took some of the British with him. The men would approach each building, keeping out of view from any windows or holes they could see. A grenade lobbed in would then be followed by a Marine, weapon to the fore, leaping into the space, scanning the interior and yelling, 'Clear!' A frightening way to fight a war: Corporal 'Nobby' Hall's section watched in horror as a dead Taliban stood upright after being shot; Marine Steve Dounias fell off an assault ladder as he attempted to break into a compound, when an RPG detonated against the supporting wall. After an intense exchange of fire, both sides hammering away at the other for a five-minute period, there was a lull, into which Sergeant Nige Quarman spoke: 'We may have wound them up a bit.' There were, though, surreal moments when training days loomed large for some of the younger marines, one of whom was heard to ask if it was OK to have a grenade to lob into a compound. The corporal alongside him, professionally practical, replied by offering his: 'Here you go, mate, use two.'

The last compound to be attacked contained a substantial weapons cache and 'improvised explosive device' factory, with all the wiring, explosives and 'cooking pots' needed to construct the bombs, confirming, if any confirmation were now needed, that the low mud and shingle buildings were – or had been – a high-priority Taliban hideout. All weapons, ammunition and equipment were eventually recovered to FOB Price for examination by the Weapons Intelligence Section.

Second Lieutenant 'Dickie' Sharp's 2 Troop was ordered to push on to the sluice gate. Forging east, Sharp's troop faced running battles with Taliban who struggled to keep up with the mobile marines while at the same time trying to avoid the unwanted attentions of the Apaches and Harriers that had been called in. The marines swept into the neighbouring alleyways and compounds to return fire. The ground opened up around here and from being fields and compounds was now thicker with trees – dry and leafless, still, but better cover all the same. The men were able to shelter alongside a high sand and shingle bank, with a shallow stream behind them. Here at the sluice gate the engineers who had accompanied the attack were able to determine that it wouldn't be feasible to build a permanent checkpoint: nobody chose to disagree. Meanwhile the enemy were regrouping and starting to crawl forward, the ever-present RPGs bursting about the marines as they did so. Corporal Bailey's Viking had nine RPGs aimed at it but none hit – some exploded harmlessly in the air while a couple hit the bank in front and skidded off to the side.

Eventually 2 Troop, surrounded on three sides, was given the order to withdraw. Corporal Thompson's WMIK, the leading vehicle, drew all the fire on to itself, so that the five armoured Vikings could pull back. With attackers on three sides, Thompson continued to pound out heavy fire to suppress enemy positions, allowing the vehicles to move off swiftly. Corporal Heath, who remained exposed throughout the contact while accurately calculating target grids, successfully controlled 81mm mortars and 105mm shells on to the enemy trenches and compounds. The B-1B bomber had the final word as the JTAC, Corporal 'Larry' Lamb, coordinated the dropping of a range of ordnance, including a dramatic finale of three 2,000-pound bombs in one 'stick' on to the stubborn enemy. The trench was silenced. One marine commented over the air, 'Nasty … but problem solved.' As the company pulled back, clearing any unattacked compounds as they did so, the Vikings provided close-up fire support for this phase while artillery 105 shells fell on those few Taliban positions still firing.

At FOB Price Corporal Cowe summed up: 'A lot of different emotions came out during the first few hours. No one showed fear at the time. The funniest thing was that even at the height of battle, everyone was shouting at each other saying how "hoofing" it was, laughing, giggling and making a joke of the situation. You may as well laugh when things get tough; you will only cry if you don't. Bootnecks are good at that and it is a great pressure relief. The bond within Juliet Company yet again shone through.'

Cowe's praise for the men of Juliet Company was matched only by his relief that all had returned, remarkably, unscathed. 'During the course of the four-hour battle, there were many individual acts of bravery. Everyone performed to the highest standards; those in command made the correct decisions; those who followed instructions did so impeccably.'

Perhaps the part of the day the men enjoyed most hearing about was when they were safely back at base. The Estonian contingent within the NATO forces cut off some Taliban reinforcements, and the ICOM chatter that was picked up afterwards was 'brilliant, along the lines of "We are here and they are surrounding us. We need reinforcements." And the reply came back, "Our reinforcement route is cut off. We cannot get to you. We will pray for you."'

'So we did not achieve everything we set out to do, although we did achieve the raid on the building we had earlier targeted,' reported Ewen Murchison, 'During the four-hour contact we used every one of our supporting arms and Apaches for about an hour and had GR7 drop bombs as well. After four hours of this we had dealt with them in three areas but they were still fighting hard in one area. My mission had not been to clear them but to destroy them. I counted myself extremely lucky and wasn't prepared to push my luck any further but when one considers the weight of fire, the proximity of the enemy and that we came away unscathed after a four-hour firefight, it can only be put down to the superior skills, training, tenacity, intelligence, courage and fieldcraft of the individual marine.'

Three days later, on 13 January, Mike Company, attempting to push the Taliban back from the Kajaki dam, were involved in more contacts, one of which was to lead to a tragic conclusion. FOB Zeebrugge was still receiving mortar fire and 107mm rocket attacks daily. Although the firing points could vary, there was one more frequently used than the rest, three kilometres north of the dam: 'Nipple Hill' had been identified as a mortar fire control position.

Mike Company moved off early one morning to remove the enemy from the hill and its surrounding compounds. As the first streaks of daylight caught the top of the hill, two enemy sentries were seen with weapons. 10 Troop were in limited cover, 300 metres away from the Taliban position. Mortar bombs were lobbed on to the trench positions at the top of the hill and the marines moved up for another 150 metres before the order to 'fix bayonets' was given. Twenty-year-old Marine Ash Hore's view was that 'fixing bayonets for the first time since training was awesome'. Marine Mathew Bispham, who was to win the Military Cross, added, 'Afterwards it was great but at the time it was honking.'

Lance Corporal Owen was in the lead vehicle at the bottom of the hill. 'As we were watching, eight blokes dressed in black robes with weapons came out. Two or three of them stopped to fire at our guys who were on foot halfway up the hill.' He whipped the GPMG off the WMIK and set it up on the ground, firing up at the Taliban. This gave the marines on the hill the chance to close in on them, pepper-potting their way up and quickly clearing the trench system. Sounds, however, came from the bunker; it was a communication radio, 'So the room was cleared with a grenade,' said Corporal Jack Scott. As the radio was still working when we entered, we gave it to our interpreter and from then on we could hear everything they said. One group was discussing our presence: "The Brits are coming ..." "How many?" "... I think they have sent all of them for us ..."'

Although the enemy had retreated from their locations, the men on the hill had no rest, for the Taliban simply took up new positions in a compound further down the hill, on the other side. With the aid of the Apaches above, firing their Hellfire missiles, and moving slowly towards each low building in turn, Bertie Kerr's 11 Troop approached its objective. As the Apaches had been unable to shift all the enemy because the position was too well fortified, it was decided to launch a direct ground assault.

Without a moment's hesitation Marines 'Vinders' Curry and Matthew Bispham – who had arrived in 42 Commando straight from training – charged forward over 50 metres of open ground with their bayonets fixed. As they neared the comparative safety of an outer wall, a single gunman leapt from his hiding place and shot Tom Curry, killing him instantly. Matthew Bispham immediately shot the assailant and, 'fighting almost hand-to-hand', as the citation for his Military Cross reads, killed a second Taliban in the compound itself.

Marine Tom Curry, who was twenty-one years and twelve days old, was described in his obituary as a 'self-effacing, utterly unselfish individual never slow to have a laugh at his own expense, who was, in typical fashion, at the front, courageously closing with the enemy with no thought for his own safety, just that of his colleagues who were close by ... a glowing example of what a Royal Marine represents: courageous, robust and highly professional ... that he carried these qualities as a young man into the dangers of battle speaks volumes and we are all immensely proud of him'. 'Vinders' Curry was to be awarded a posthumous Mention in Despatches.

Having again pushed the Taliban back and after finding numerous weapon caches, Mike Company withdrew to the dam, carrying with them a badly wounded Taliban fighter. This enemy soldier later died but not after a British medical team had tried everything to keep him alive. When asked, pointlessly, by a journalist, 'Why did you try and save him?' the marines replied, 'Because that's what makes us different from them.'

The attack on Nipple Hill was a success, proven by the lack of incoming mortar fire for two weeks, and even after that the frequency and accuracy were greatly reduced. The deduction was that among those killed was the Pakistani or Pakistani-trained operator – usually the only Taliban with the correct skills.

Further south, the attack on Jugroom Fort, a formidable objective about five miles away in the green zone, was imminent. India Company's 1 Troop was moved into Garmsir to protect the District Centre while Zulu Company moved south for the operation. In order to mislead the Taliban into believing a large-scale assault was being mounted from the north, 1 Troop was also tasked to attack four objectives simultaneously; this would, it was hoped, divert attention from Zulu Company's southern approach.

Corporal Pete Harvey, of 1 Section, was briefed to take his marines one and a half kilometres south of the DC to an objective nicknamed Snowdon and, on H Hour, to fire two ILAWs into a suspected sentry position. Once that was done he was to fire every weapon he had at his disposal into the adjacent enemy compounds. 'En route,' he was told, 'you'll need to clear Strip Wood.' The wood was known to be an enemy firing point, with a recently discovered trench system.

In near darkness on 14 January Pete Harvey's section led the troop southwards before breaking off and moving along the river bank to complete its own tasks. Making steady progress, while continually searching the ground ahead and to the flanks with a thermal imaging camera, they soon arrived at Strip Wood. Once they had cleared it without incident, the section continued its stealthy advance until an identified heat source was detected 200 metres to their front. The men sank to the ground, to hold their position as quietly as they could while two snipers were sent ahead. The minutes stretched out painfully as the cold night air engulfed the marines, until a pair of muffled bangs told them that the 'heat source' had been taken out with two .338 rounds. The marines again started moving slowly south through the wood until the thermal imaging camera told them they were only fifty metres from two Taliban in a well-defended bunker. Corporal Harvey and Marine Cain moved to observe but backed off silently as one of the occupants left their hide. The section commander now moved his men to the river bank to check if they could 'right-flank it', but as it was so well concealed it would have been impossible to approach. Pete Harvey now received a radio message stating that this was his objective. He positioned his fire support and ILAW team accordingly and waited for H Hour.

Again the time dragged, whether as a result of the cold or the tension no one knew – or frankly cared. Finally H Hour did arrive and Harvey tapped the two men with the ILAWs to go ahead and shoot. He moved swiftly away as the back-blast from the ILAW is almost as dangerous as what comes out the front. The first missile, fired by Lance Corporal Woods, was a direct hit, going straight through the window. The second, shouldered by Marine Cain, was less successful – 'It was last seen heading for Pakistan,' said Harvey delightedly. The Taliban in positions way to the east side immediately woke up to the fact the British had come behind them and launched an attack, firing haphazardly in the dark at what they thought were their targets. The green Taliban tracer, though, was close enough to keep the marines back in Strip Wood while they called in artillery and mortars. Once the Taliban were suppressed the marines moved in to check the damage on the sentry post.

'The bunker was well constructed and we had done well to defeat it in the initial contact, collapsing the roof and killing the occupants,' said Corporal Harvey. 'Our mission had been a classic, small commando raid, conducted by a group of determined men who moved – unseen – into the enemy's rear, unleashed hell and extracted back to fight another day. Only when we arrived at FOB Delhi could we relax, discuss the night's operation and listen to the sound of Zulu Company's attack further south.'

Meanwhile – and, as it was to turn out, fortuitously – Colonel Magowan's Headquarters was due for one of the many personnel changes forced upon it by its dual role as an under-staffed manoeuvre battle group and brigade command support group. Major Andy Lock had arrived in theatre to replace Major Sean Brady as the group's operations officer, bringing with him kit for just three days: he was to remain in the field for nearly four months.

By the time Lock arrived at the desert base, the Jugroom Fort assault planning staff were 70 per cent into the battle preparation. Earlier, Lock and Brady had been working side by side in Fleet Headquarters, Portsmouth, when a call came for volunteers to augment the brigade; they both put their hands up, so it was decided that each would have the opportunity to serve for half a tour, with Major Brady deploying first.

Andy Lock had joined the corps in 1990 and thoroughly enjoyed his time as a marine: 'I had eighteen months in 40 Commando; the easiest and most comfortable years I ever had.' Nevertheless, he spurned that 'comfortable' existence, applied for and was granted a commission in 1993; joining the same officer intake – or 'batch' in Royal Marines parlance – as Sean Brady.

Now, in the western desert, and with mounting concern – he was not alone – he had watched the events at Jugroom unfold and so was not surprised when, on everyone's return to Garmsir's Agricultural College, he was invited to assume command of Zulu Company with immediate effect. Although no stranger to 45 Commando – his previous command had been Yankee Company – he was surprised that a 45 Commando officer had not been chosen for this awkward and sensitive task. The reason, though, was clear: all that commando's officers were deeply involved in some of the most complicated and useful work of the brigade's deployment: the unsung duties of the Operational and Mentoring and Liaison Teams.

On assuming command of Zulu Company, Andy Lock had two vital and delicate tasks to oversee and perform: the collection of Marine Jonathan Wigley's kit and the first meeting with the officers and senior and junior non-commissioned officers. As is always the case within the Royal Marines, the eventual kit sale – held after the deployment – was impressive, 'really poignant' and a fine tribute to Marine Wigley. Financial common sense plays no part in these events, for the only aim is to raise as much money for the deceased's next of kin as is possible, with some colleagues known to pay £100 for a box of matches.

The new company commander's second duty was to face his men. He claims to remember little of this delicate meeting in Garmsir's District Centre, other than the words of the company sergeant major afterwards: 'The best thing you said, sir, was all that you could offer them was leadership. That's all they wanted to hear.' Coming from the 'remarkable' CSM Shepherd, that was support indeed. They were to make a first-rate team.

Operations across Helmand did not reduce in intensity during the latter half of the month.

At Kajaki, Mike Company focused on supporting the Afghan National Security Force and manning their new PVCP before handing it over completely. Martin Collin placed some of his troops with the ANSF's OPs and in the police station, all of this with a view to building ANSF confidence and thus their capacity, and ability, to take on more security tasks within the area. This posting was a success, apart from the Afghan tendency, as the men saw it, to bake and slaughter animals every day.

Of the initial set of targets identified by Mike Company during its early observation phase at Kajaki, one was the village of Barikju, a collection of low compounds, abandoned by the locals, that lies off the eastern edge of the M1 wadi, three kilometres north of Sparrowhawk. It was an identified firing point and a suspected command, control and administration base, and the northern village of three that included Lower and Upper Khalawak. Captain Tony Forshaw, the company second-in-command, unknown to his commander, began hatching an ambitious plan to assault all twenty-five enemy-held compounds.

The company commander, Major Martin Collin, was due on his R-and-R leave on 7 January, but he had delayed it by a week to supervise a series of small operations that steadily expanded his company's footprint and influence. He finally felt able to fly home on 14 January, but when it came to the day he was reluctant to go as only the day before the much-admired Tom Curry had been killed. Conscious of the effects not only of Curry's death upon the other men, but also of the exhausting times they'd had after carrying out three or four punishing company-level operations, Collin told Forshaw to give the lads a breather for the fortnight he was away: 'Just conduct a few recce patrols, build up a picture of where the Taliban now are.' 'His last instructions to me were, "Plan for more ops on my return," but the moment he was on the Chinook I was behind the typewriter tapping out my Concept of Operations for an attack on Barikju.'

As Forshaw needed intelligence for his planning processes he fortified a position known as Kidney Bean Hill on a ridge to the north of the river. From here, at sunset, the Taliban who had walked down from Barikju could be observed taking up firing positions in Khalawak. Although from Kidney Bean Hill the enemy's approach was hidden, the moment they were in the village and attempting an RPG shoot they could be seen and engaged. The Royal Marines on the recently-occupied Sparrowhawk OP would also be waiting, for they could cover some of the enemy's routes. Once the Taliban were in the narrow and restricted alleyways, the OP above FOB Zeebrugge opened up with their long-range weapons; the heavy machine guns and Javelin. Unusually, for they were credited with more professionalism, the Taliban carried out this set routine seven times – each time taking casualties in exactly the same places.

Forshaw decided that this tedious routine needed to be stopped once and for all, and despatched a night patrol to lay Claymore and bar mines in the alleyways and bunkers into which the enemy jumped each time the marines started firing at them. 'And when we saw eight or ten coming in we set off one Claymore and then another bar mine as they started to withdraw.'

But the Taliban became wise to the Royal Marines' tactics. Forshaw had established patrols on Khalawak Ridge, and, after two or three small ambushes of the Taliban 'carried out a proper big one. When we had stopped firing, the guys on Sparrowhawk called and said, "Boss, get the hell out of there, *fast*!" For about ten kilometres out to the north all they could see were lights of mopeds, motorbikes and cars all converging on our position on the ridge, eighty or ninety vehicles approaching us. Just me and eight lads – and fairly low on ammunition. So we packed up the three WMIKS and drove back into the main camp.'

After Forshaw and his men had left the ridge, the OP at Sparrowhawk continued observing. The vehicles hurtling in from the north-west and north-east congregated on the hill, headlamps blazing 'as the whole place just lit up'. The ambush was clearly pre-planned – but totally unsuccessful.

Now that he had enough intelligence, Tony Forshaw continued with his planning for *Operation Volcano*. Intelligence, thanks to the use of a Desert Hawk, indicated that a position even further to the north was being used as an arms cache. Another OP in use was on Ant Hill, which was riddled with interconnecting caves, holes and tunnels. Sentries were posted, changing at regular intervals and doubling up when noises were heard. Groups of twenty or thirty men could be seen emerging from bunkers from time to time. It was clearly a substantial base in the area.

The idea was to clear all the Taliban in Barikju. 'I knew Upper and Lower Khalawak were deserted, as a patrol had found a tethered dog trained to bark as an early-warning system. A second patrol left some poisoned meat and that dealt with that little obstruction.' As a pre-operation precaution Tony Forshaw asked a B-1B bomber to drop three 2,000-pounders on Ant Hill – which they did with their usual devastating effect.

The operation was then delayed, twice. The first time was because the mission was to take place in the afternoon, while avoiding the two villages of Upper and Lower Khalawak, but by then a low mist had settled and without over-watch Forshaw decided, sensibly, that it might prove too risky. Colonel Matt Holmes put the operation off the second time, for neither a Nimrod nor a B-1B bomber was available. Therefore the attack started on 30 January, the day Company Commander Martin Collin returned from leave. Forshaw's description of the moment Major Collin realised his men were not taking things relatively easy, as he'd suggested, but were actively engaged, is dry: 'The boss landed at Camp Bastion and went into the Joint Operations Centre to see the Nimrod down-feed on the big screen and he watched these chaps running across open ground and attacking a village. Only then was he informed that that was his company conducting a full-blown attack on to Barikju. I think he was quite pleased, but at the same time he was quite peeved that he was out of it. He had a fairly good idea that I wasn't going to miss out on an opportunity.'

Experience had shown that each time Mike Company attacked a village the Taliban ran out of the rear to lose themselves in the wide M1 wadi to the west. This is a flat, usually dry water course, anywhere between 100 to 150 metres wide, with a bed of sand, shingle and rubble; when it rained, the nature of the ground caused dangerous flash floods that quickly turned into an even more hazardous torrent. Forshaw ordered the OP on Sparrowhawk to watch for the possibility of 'leakers', as they were known, while he covered the northern end from a place dubbed Blue Pipe Compound; here he established a GPMG gun line with three or four guns. Shooting at targets from a distance wasn't a problem, with a .5-calibre sniper rifle up there: 'We used it against an armed Taliban we had positively identified on a motorbike and at 1500 metres he got an armour-piercing round straight through the back. There wasn't much left of him.'

Additionally, Apaches would be held back ready to engage any retreating or escaping Taliban. They would stay out of earshot and only at H Hour be brought to a point five minutes away.

As on previous operations, it was important that there was 100 per cent identification of the target area before any bombing took place. Forshaw led the company out to Khalawak in the middle of the night and holed up until first light, before moving off to start the operation. Identification was easy – the sentry took a potshot at them with an RPG – so the marines opened up with their mortars. At this stage they were about 150 metres away from their first target, Compound 1.

'The Nimrod had already seen a build-up of seven or eight Taliban in the compound and a different one they had moved to, that we called Compound 10. There were further enemy in 8, 9, 14 and 22 Compounds.'

Certain now of his target, Forshaw called in the bomber to drop a 1,000-pound bomb. The buildings of the compound were totally levelled by the explosion and the dust cloud that rose up covered the marines' approach as they ran across the open ground. Holding the high ground gave the marines an advantage and one that Forshaw used happily: 'I put myself at the top end of Khalawak plus a Recce Troop gun team with Minimis and snipers who engaged any Taliban that fired at us from Barikju. They could also clobber any that tried to escape around the top.'

The three troops Forshaw used then advanced in staggered formation through the compounds, although the bombs dropped had created such devastation there was very little for them to do. 'They were still digging out bodies five days later, when the Taliban from Chinah made the locals go in and get them out, which can't have amused them very much.' The mortars were less effective; the walls of the compounds being so thick, nearly eight feet in places, and the roofs two feet thick – that unless they managed to catch a Taliban out in the open there was little the mortar could do, 'although they made a nice bang'.

Thanks to the bombing and speed with which the marines worked their way through the compounds, the operation went better than expected and 'we got all the way to the Line of Exploitation. I didn't want to go any further than that or we would get exposed to the back edge of Pyramid Hill where there would be Taliban.

'Also during the week we had watched some enemy moving stores – weapon bundles and that sort of thing – about 2km further up the wadi, so they probably reckoned that something would be happening soon. We had dropped quite a few bombs that had pretty much buried all the Taliban that were there – so I guess they are still there. That was very much it, except that there was activity on Ant Hill, which is why we dropped a bomb on that too.'

Meanwhile the Apaches got four or five who were creeping

out along the edge of the wadi. Forshaw waited till everyone else was clear before pulling out himself, and to make quite sure the underground passages could not be used again decided to drop one last bomb on Compound 1 as they had found deep tunnels and an arms cache there when running through it. 'And that was pretty much that; Marty pitched up on a flight that afternoon to say, "Well done."'

Having completed their mission and taken no casualties, the marines regrouped and immediately patrolled out of Barikju towards Kajaki to await further tasks.

6
FEBRUARY:
ADVANCE TO AMBUSH

'...not seen in the photograph are the...hordes of marines behind me, so there was little chance of me being taken hostage; my men would not have let me go alone any distance at all. There was a Canadian who went to such a meeting and he was chopped in the back of the head with an axe. Whether true or not, it focused the mind when you sat down with these people.'

During February, Colonel Matt Holmes continued to move his companies about, in order to rest some, while at the same time building up troops in one area – Kajaki – in readiness for a major assault. Lima Company moved from Camp Bastion to Now Zad, where they would stay until the end of *Herrick 5*, being relieved only at the very end to take part in *Operation Silver* at Sangin in April. Kilo left Now Zad to join Mike Company for *Operation Kryptonite* at Kajaki, on completion of which Mike Company moved to Camp Bastion for a brief respite before replacing Charlie Company of The Rifles at Sangin: a hard and final sojourn that would cost them dear. Juliet remained 'MOGing' out of Gereshk's FOB Price.

Tony Forshaw's clearance of the Taliban from Kajaki's neighbouring compounds showed that success against the enemy in their own territory was possible. As the attacks against FOB Zeebrugge and its observation posts continued to delay and disrupt work on the dam, Colonel Holmes knew that a brief overlap of two companies at Kajaki would give him the ideal opportunity to conduct a major assault against established enemy positions, thus solving the problem once and for all. He felt he had sufficient momentum behind him to clear Chinah and Shomali Gulbah after breaking in at a location known as Banana Lay-by: a bend in a track from where the marines had taken considerable Taliban fire. He also wanted to expunge the Taliban from Pyramid Hill, a little further to the north.

On 11 February Kilo Company and 42 Commando's TAC HQ flew into Kajaki prepared, with the resident company, to clear Taliban from the areas of Shomali Ghulbah and Chinah. Kilo Company arrived in Kajaki with two Close Combat Troops, one Fire Support group, an ISTAR group of about forty-five ranks, one Royal Engineer troop, four mortar barrels, a Fire Support team, a Desert Hawk detachment, four 'medics' and a Y Squadron Electronic Warfare team. Mike Company was sharing FOB Zeebrugge with an OMLT; furthermore, ANP, ANAP and militia were manning a number of permanent positions circling both the dam and the FOB.

Colonel Holmes established his command post for *Operation Kryptonite* at the site of the cave systems on Shrine Hill. When the operation was completed, the bunkers were then mined, to make sure 'they could never be used again'.

The attack was planned to start at 0615 on 11 February, but in the very early hours of the morning 11 Troop, under Second Lieutenant Bertie Kerr, crept up to the compound walls and as quietly as possible placed mouse-hole charges, ready to blow when H Hour came so that they could break into the compound as the first explosions rocked the Taliban.

The original mouse-hole charge, made up of five small charges on a wooden cross, had been used early in the deployment by the sappers of 59 Independent Commando Squadron, Royal Engineers, but had proved too large and cumbersome to carry between compounds, especially when under fire. After trials it was decided that one single, slightly larger, charge that could fit into a day-pack would do the trick.

Kerr was looking intently at his watch and about to signal for the charge to be detonated when the attack was suddenly called off. At that precise moment – and so just in time – the colonel received a radio message from Camp Bastion cancelling the expected Apache support and any chance of a Chinook for casevac: fog at the camp was preventing all flying. Kerr took his troop away from the wall and returned to the FOB, hoping almost as fervently on the way out as they had on the way in that they would not be spotted by the Taliban. They were lucky that their attempt on the wall had not been discovered: the attack plan had not been compromised; overwatch of the area from the three OPs indicated a normal pattern of life that continued throughout the day.

Another Rehearsal of Concept was held that afternoon and, with new timings reaffirmed, all was ready for the next day. To while away those anxious hours before a set-piece assault the marines of the two companies played an impromptu but mildly serious game of cricket which, for their colonel, was 'quite a poignant moment, watching my men relaxing knowing that they were about to go into battle. They all knew the risks.'

The following morning saw the marines form up in a freezing-cold pre-dawn, across a field below the bombed-out shrine and cave complex, waiting for the first of 11 Troop's charges to detonate and thus signal the start of their day's work.

11 Troop had repeated their clandestine approach and were ready and in position by the Banana Lay-by compound before dawn. The colonel patiently waited for morning prayers, blasted over the village tannoy system, to finish before giving the order. On time, 11 Troop exploded their way into the compound with their

mouse-hole charge, as the first of 42 Commando's mortar bombs landed to the rear of the Taliban, blocking their exit route. In the distance, the menacing – or reassuring, depending which side you were on – sound of Apache helicopters replaced the mullah's lilt: the aircraft were starting their thermal-imaging search for enemy movement. With mortars landing around them and marines pouring through a sudden gap in their wall towards them, the enemy knew that they were facing a hard, intelligent fight and one that was to last for the next sixteen hours.

The commanders of the two companies were old friends, able to read each other's intentions without too much radio chatter. Neil Sutherland's Kilo Company was tasked with clearing the village of Chinah, a maze of small buildings, compounds and rabbit warrens to the west of the M1 Wadi, while Martin Collin's Mike Company was to clear Shomali Gulbah, a similar village one kilometre to the west and across a smaller wadi – the M6. Once both villages were secure, Kilo would continue the push northwards to take the heavily defended Pyramid Hill, north of Chinah.

Clearing compounds was becoming something the marines felt they had mastered. With the helicopters above, as well as the OP positions on the high points around them, they could confidently focus on each compound as they went through, knowing that someone would tell them if reinforcements were being brought up. The same practice – using grenades to clear their path as they went into each new compound and its buildings – was employed. The work was dangerous and delicate, as every wall might conceal a firing point, every dark spot a waiting gunman; so they moved slowly through each one. They found after a few hours, though, that most of the compounds were simply abandoned by the Taliban, scattering before the marines as they steadily made their way forward. The enemy managed only sporadic small-arms and RPG fire, with no coordination as they fell back.

With Mike Company pushing through their objective, the

Corporal John Thompson on the right with Marine Nathan Beagles driving and Marine Lewi Moncaster as top gunner in a WMIK.

M1 Highway with convoy.

Juliet Company, 42 Commando, on a Mobile Operations Group operation.

WMIK versus anti-tank mine on the approaches to Sangin for Operation Silver.

CSM Marty Pelling's Viking immediately after a mine-strike.

JTAC Hill, Garmsir.

Sangin District Centre's vulnerable and exposed roof and tower.

Kajaki dam.

Fix bayonets! Captain William Mackenzie-Green prior to assault during Operation Volcano. Sergeant Pete Royston in the foreground.

Calling in air strikes at Now Zad.

Mike Company, 42 Commando, on a Mobile Operations Group operation.

Two inch mortar in action.

Clearing enemy-held compounds during Operation Volcano, Kajaki.

A Happy CO 42 Command (on left) returning from operations at Kajaki a few seconds before 107 rockets fell, one hitting the building in the rear.

Corporals Tug Wilson and Mick Cowe after 36 hours in contact.

Handover in the field of 42 Commando RSMs. Marc Wicks (left) to Si Brooks (right).

Corporal Al Hewett and his section leave Kajaki.

Wounded Taliban.

Farewell to Marine Jonathan Holland at Camp Bastion.

enemy spread out from Shomali Gulbah village, only to be targeted by the waiting helicopters. As the marines continued to clear the surrounding compounds, Kilo Company passed through the now-empty positions, to launch their assault on to Chinah and, with that task completed, on to Pyramid Hill and the limit of their exploitation. First 4 Troop moved towards its objective under cover of 51mm mortar smoke, then 5 Troop passed through to break into their own compounds; again under cover of heavy mortar smoke. The 51mm mortar was preferred, for, although it was being phased out, it was easier to lay down rounds more quickly than its larger cousin. It could even be carried as a personal weapon as opposed to requiring a crew of three.

Throughout the long day it became increasingly clear that the enemy, sensing an increase in British troops, had decided not to hang around – all but a stubborn, suicidal few had left. This unusual reluctance to fight would change later in the tour.

During the extraction from this successful – if not wholly satis-factory – operation, thanks to a reduction in enemy numbers, the heavens opened for about an hour. The dry, wide waterway imme-diately filled with a flash flood; a tidal wave that coursed down the M1 wadi, leaving the WMIKs stranded, unable to cross the spate. Yet the heavy weapons were needed back at the FOB, so, 'in typi-cal marine style' these and their ammunition were carried across the torrent; the marines, after a sixteen-hour contact, showing their resilience and strength as they waded back and forth until all equip-ment was saved. The WMIKs were left with their crews clutching just personnel weapons; but in full view from the OP positions, who could guard them until the waters subsided.

Settling in to FOB Zeebrugge, Kilo Company took stock of its new home, comparing it favourably with Now Zad – it even boasted beds and showers, whereas Now Zad's mud huts had neither. The governor's old summer house was not – nor had it ever been – a palace, but it was comfortable. The same could not be said for the

OP positions on Athens, Normandy and Sparrowhawk but, unlike Now Zad's ANP Hill, at least these inhabitants could occasionally come down the hill, certain to enjoy a passing comfort or two. Things at Now Zad's ANP Hill worsened during some of the daylight strikes that Lima Company's 9 Troop, the Fire Support Troop, endured. One RPG landed close to the heads, showering a marine, who was 'sitting' at the time, with the contents; a bad experience under any circumstance but with no washing facilities for weeks on end this was a near disaster for all on the hill.

The Taliban may not have been 'out in strength' for *Operation Kryptonite*, but immediately afterwards they replied with fourteen 107mm rockets into the Zeebrugge compound, fortunately without causing casualties. Launched from a possible six kilometres away it was difficult to know from whence these vicious missiles came: the OPs were targeted rather more often.

Sutherland had his full company with him plus a complete ISTAR group, in all between 180 and 200 men under command. Kajaki was an important place and worth this size of defence, for so much depended on bringing the power station up to full running order. Additionally there was a sizeable detachment of the ANA, within the compound but in a separate part. The ANA were not entirely trusted by the Royal Marines: Neil Sutherland said he 'certainly would not have entrusted our security to them if I hadn't had my own OPs on the features'.

Sutherland's pleasure in holding the high points around Kajaki – the 'simple beauty of the military situation' – was because the OPs gave them a 360-degee view, covering 10km in every direction, meaning that the Marines 'owned all the key terrain, all the vital ground'. Any potential Taliban attack could be spotted well in advance, while the OPs gave the men an excellent position from which mortars, as well as sniper fire, could target the Taliban in the compounds and alleyways below.

'Additionally, unlike most other company locations, we faced a clear "forward line of enemy troops". Simple but vital indicators, such as the changing of sentries and the movement in and out of compounds, made it much easier at Zeebrugge to keep tabs on the Taliban and the intelligence picture up-to-date. Unlike Now Zad, every time at Kajaki that I went out on an attack or an ambush I was taking with me intelligence that we had gathered ourselves. I wasn't relying on second-hand information, especially so as, from the top, we could see the whole of our area.'

Nevertheless, and despite the comparative comfort, there was no relaxation. Unlike Now Zad, where the locals had left the immediate area but still 'engaged' with the town, in Kajaki all civilians had gone. This meant the Taliban could be attacked with a certain amount of impunity: the overriding consideration being that the buffer zone had to be maintained no matter what the cost. Snipers were stationed on the OPs and were able to engage their targets across substantial distances; the furthest a sniper achieved a kill was 1,200 metres.

Sutherland used the intelligence he gained to launch raids – ambushes, as he called them – on those compounds where they could see the Taliban trying to creep back in. 'So the beauty of creating our own intelligence was that we could go out and clear compounds. We set many ambushes to make sure that the compounds we had cleared earlier stayed clear. Without enough men we could never physically hold the ground we had taken.

'Sometimes the Taliban would react, then it was the extraction that was more difficult than the insertion. What we tried to do was extract once under cover of smoke, then the next time we would lay smoke in a different place and they assumed that was where we were extracting through and so we set up an ambush and got them! When we killed them we left them lying on the ground for they would remove their own dead – always. We could have stayed and ambushed those that came for the bodies but once we had sprung an ambush we did not want to linger in the area. We stirred up a hornets' nest every time and if we had achieved our aim there was not much military advantage in staying around.'

The 'nest' had indeed been seriously stirred, proved by the numbers of Taliban now swarming towards Kajaki's FOB Zeebrugge and the OPs. With the Taliban showing no thought for their own lives, nor indeed the normal conventions of war – a situation not taught at Britain's military training establishments – it was difficult for the Royal Marines to know how best to meet this suicidal threat. Civilised nations fight in accord with internationally accepted conventions. 'We worry about casualties and the removal of our dead and wounded to the point that with the small sections we had if one man was down, then the whole section was involved in recovery.' If the Taliban had a man dead he would be left while another simply took his place: there were shades of the Korean War and the 'hordes of Chinese that came in waves'.

A real fear for Brigadier Jerry Thomas's men was the thought of capture: indeed it was 'not considered an option'. If it crossed anybody's mind it was in terms of torture prior to a beheading that would be filmed and then screened by the world's less fussy television stations.

With this in mind, and before every patrol, Neil Sutherland used to consider how far the patrols could push out, weighing up the numbers of men involved, the nature of the vehicles they travelled in or which accompanied them, and the limitations of those vehicles coupled with the casualty recovery procedures that required open lines back to the FOB or access to casevac. 'We had established their "front line" and so forced them away continually in all directions: keeping them from the dam. We probed in and out the whole time and in doing so maintained confusion in the Taliban minds: "dynamic unpredictability" was the brigadier's requirement and that is what we were achieving.'

The overall aim was not necessarily to kill Taliban but to ensure that the dam's workforce was safe and ready to receive a new turbine from China. This meant guarding the thirty to fifty local workers as they moved in and out, although the Taliban let them through for it was in their interest that the power station worked. They needed the electricity to flow so they could charge for it illegally by extortion, although they had other ploys, such as cutting down telephone poles to make the locals pay for their resurrection. To do so meant the enemy had to get close enough to conduct their nefarious affairs and to do that they had to attack the British.

In the end, though, Sutherland knew that the terrorist can be removed only by the indigenous population and not by the military. In Helmand the problem was sifting out those who were reconcilable from the irreconcilable ones and, to add to this conundrum, there were different levels of Taliban, including the foreign fighters who, some believed, were there only for the money.

Sutherland would try to discuss these aspects of his duty with local Afghanis to the west and south. Twice he held a *shura*, once with the elders of Shebaz Kheyl and once with those in Kajaki Oyla, but each time he found them wary: the earlier fighting 'had been significant and they did not want to be caught up in any more if they returned'. His battles with the Taliban were, perhaps, rather easier than his battle for the hearts and minds of the locals. At Now Zad he had made considerable progress but in Kajaki any such rapprochement remained a frustratingly long time away.

Elsewhere across Helmand, operations to track down the Taliban continued unabated. On the morning of 13 February Ewen Murchison's Juliet Company conducted *Operation Bauxite*, a patrol fifty kilometres or so to the north of Gereshk and close to Kajaki in the upper Sangin Valley. The target that day was Kajaki Sofla with the intention of disrupting Taliban activity while at the same time observing the pattern of life so that an intelligence picture could be built. Murchison split his company into two groups, each operating either side of the wide wadi that runs away to the south of the village.

CSM Marty Pelling was on the western side of the wadi with the second-in-command, Captain Bruce Anderson, plus the mortar line. They were escorted by 1 Troop's three Vikings, commanded by Second Lieutenant Sam Hughes with his troop sergeant, Willie Whitefield. Half of 3 Troop with Second Lieutenant Duncan Law and Sergeant Gaz Watford provided the Fire Support group from their two WMIKs.

They were to drive through territory that had once been held by the Russians years before, with signs of their occupation still visible among the desolate rolling dunes. Hull-down positions for armoured vehicles along with shell scrapes for individual protection and observation marked their route. The patrol's way through was along a narrow ridge where, as they encountered signs of an earlier war, the company's assault engineer, Sergeant Jim Tyler, reminded the CSM of the likelihood of a legacy mine threat and suggested that this practical warning be passed over the company radio net.

The men squashed sardine-style into the back of the stuffy, dim, rear Viking cab listened intently as Captain Anderson relayed this grim message. No one knew what a mine would do to a Viking, how it would hold up. Shortly afterwards – almost as if the Taliban had been listening to the Royal Marines' ICOM chatter themselves – the team heard the sound of small-arms fire and, as they de-bussed from their armoured vehicle into bright Afghan daylight, suddenly found themselves 'in contact'. The shriek of numerous RPG rounds, the *crump* and *thud* of mortar fire pinned them down. A red streak flew overhead – a rocket – and shot across the top of the Viking, exploding harmlessly in the desert sand beyond.

On the eastern side of the wadi, the other half of the company also ran into Taliban forces. The marines fired their own mortars in retaliation and sought targets for their more sophisticated weapons, but the Taliban had embedded themselves successfully into positions where they were hard to spot, while the British were mostly in the open.

With the initial contact quickly developing into a full-blown firefight that was to last over two hours, the CSM, monitoring each troop's ammunition scales, realised that by early afternoon it was necessary for him, and the company quartermaster sergeant's Viking, which was carrying the majority of the reserve ammunition, to move forward with a resupply. Pelling's Viking carried small quantities of ammunition, but it did contain two medics, Leading Medical Assistant Al McNeil and Marine 'Chocolate' Coleman, along with Sergeant Jim Tyler and Marine Al Hayward, both Assault Engineers.

Their first logistic run, up a small re-entrant in dead ground, was delivered to 1 Troop, now seriously low on small-arms rounds. Dropping off Willie Whitefield and Corporal Tug Wilson, along with a working party, the two Vikings then pulled back to high ground where they met Gaz Watford and half of 3 Troop. About 400 metres to the east, 3 Troop was enthusiastically engaging the enemy with .5-calibre and Javelin, and were desperate for a resupply. Half the Viking's .5-calibre load was delivered along with the remaining anti-tank missiles and a few boxes of 7.62mm link. Now it was necessary to resupply the other half of the company; a route was chosen off the high ground towards the wadi. They didn't get very far.

When the CSM's Viking left the ridge to the south with the CQMS's vehicle following, Marty Pelling had his head down, staring at the map that he was using to direct his driver, Marine 'Robbo' Robertson, across the dry watercourse. As the Viking accelerated there was a loud explosion – although muffled by headsets – and everything went black. The vehicle jumped forwards, then lurched to a halt.

'Fucking hell!' yelled CSM Pelling. The cabin rapidly filled with smoke and dust, 'like being inside a blender with the contents of a vacuum-cleaner bag'. He could hear Robbo calling for help, and he knew from the sound of the man's voice that he was injured. He himself felt no injuries. Two immediate thoughts ran through the sergeant major's mind; it must have been an RPG strike, and second, 'Please let the door open, don't let me be trapped.' As the smoke thickened he thought he could feel the heat of a naked flame and prayed that that he wouldn't burn to death.

He tried the door; it wouldn't open. He tried a second time; still no luck. Third and fourth time and, with the heat rising up round him, the seal cracked and the right-hand-side armoured hatch swung up. Pelling clambered out and realised why it had been tough to budge in the first place, for there was no obvious mechanical problem. With the loss of the left running gear the Viking had come to rest at an angle; with the heavy passenger door facing upwards it needed more force than usual to open.

The ominous sight that greeted Marty Pelling was red fluid pouring out between the tracks. Horrified, he thought, 'Oh, my God, that's Robbo; he's been torn apart.' The rush of air into the cabin caused the flames to leap up, but with the smoke clearing from his lungs Pelling was able to react calmly. He reached back into the Viking, scrabbled under his seat, found the fire extinguisher and swiftly doused the fire while Marine Al Hunter, who had been sitting behind Pelling, was trying to move Robbo from the front seat. Sergeant Jim Tyler, the Assault Engineer, realizing immediately that his Viking had hit a mine, was yelling at Hunter not to move out of the vehicle but to climb out through the top hatch. Marine Mikey Dunstan, who had been on top cover, had sustained a lower-limb injury; Tyler was already dressing his wound.

Now that Robbo was being helped both from inside and outside, Pelling recovered his rifle and bag containing maps, notebook with the company 'nominal' – the Troop Bible – and took stock. As he did so, the CQMS's Viking drew alongside. Anxious to send an immediate Sitrep the CSM was pulling on his headset, only to be met by silence. His radio, too, had been destroyed but using the CQMS's radio he managed to pass his message to the OC.

The mine had detonated beneath the centre of the left-hand track, below the driver's seat. The force of the blast had picked the Viking up and thrown it forward, so that the mine crater was now a few feet behind the 13.5-ton vehicle. The armoured rear door, behind the driver, was hanging by one hinge with its entire bar armour 'cage' somewhere across the desert. 'If anyone had been sitting behind Robbo they would have been killed,' said Pelling. 'Incredibly, Marine Al Hunter was sitting behind me and walked away uninjured: testimony to the strength and build quality of the Viking which, that day, saved our lives.'

Having sent his Sitrep, Pelling returned to his damaged vehicle to assist those helping 'Robbo'. 'Chocolate' Coleman was already applying first aid to Marine Robertson, who was hanging out of the door, which had been blown partially open, but he was still trapped, and unable to move his legs. The only visual injury was a cut on his nose; the 'red stuff' Marty Pelling had seen pouring from under the Viking was hydraulic fluid. Pelling asked Robbo how he was doing, what he could feel in his legs. The driver looked down at his trapped limbs, and at the wreckage strewn around from the damaged Viking: 'Take a picture, Sarge; no one will believe I didn't get blown up in this.'

A rifle section from 1 Troop led by Corporal Mick Cowe appeared, running along the ridge before descending to assist. Sergeant Tyler and the others on top of the Viking shouted loudly, 'Minefield!' – but this didn't stop the men – 'At which they slowed down a bit, but to their credit they just bashed on over to help out.' Small-arms fire suddenly crackled out, the bullets whistling past the stranded machine: Corporal Cowe lined his marines along the ridge to return fire.

The driver's side of the damaged Viking had been badly buck-led in the blast and the bar armour so twisted that the door could not now be opened. To begin with, Marty Pelling thought that the best way to release Robbo was to dismantle the remaining bar armour, so a torque wrench was found, but with grease and sweat covering their hands it slipped and hit Robbo in the face. 'As if things weren't bad enough for him he displayed that crucial commando quality of "cheerfulness under adversity". He simply looked at his crewman and said quietly, "Crate" – meaning that hitting him in the face would cost a crate of beer.'

A few more minutes of futile 'spannering' ended in the realization that Robbo would have to be pulled, physically, from the wreckage, a task that Marine Coleman was loath to attempt for fear of injuring his friend further. Either way it was obvious that a casualty evacuation Chinook was needed urgently. Lieutenant Hughes confirmed the details with the company second-in-command and the request was passed. The other casualty, Marine Mikey Dunstan, was being treated by Al McNeil, who had placed him on a stretcher with his 'patient' looking 'fairly happy as the morphine kicked in'. Now it was a question of taking two heavy, injured marines the couple of kilometres to the extraction point, away to the south and in dead ground.

Throughout this the remainder of the company was continuing to fight an increasingly fierce battle with an enemy who, having witnessed the mine strike, were trying to exploit it to their advantage. The marines who had been rearmed during the resupply were popping off Javelin and .5-calibre, forcing the Taliban to keep low and preventing them taking that advantage. The light, though, was fading as the afternoon slowly drew to a close and time, to some extent, was running short.

The helicopter landing site was south of the mortar line, who were able to provide over-watch. The moment the Chinook was on the ground the on-board 'immediate reaction team' disembarked and fanned out into all-round defence; the medical team followed. With the two stretchers loaded the Chinook was gone, the pilot not wishing to hang around too long at the wrong end of the Sangin Valley.

With the casualties away, Jim Tyler prepared the Viking for denial. Stripping out what he could and, ready to use a bar mine to act as the main charge, he opened up the jerry cans attached to the top of the rear cab and wound detonating cord over them. Once the Viking had been denied the company could begin to withdraw, dangerously low on ammunition and in semi-darkness. As Juliet pulled back through the helicopter landing site the charges on the Viking blew, sending sparks flying hundreds of metres into the air and – the marines hoped – scorching any Taliban who'd come too close. The company sergeant major conducted a hasty 'ammo-redistribution', while tallying up the totals of spent munitions, which included eleven Javelin missiles and 18,000 rounds of 7.62. It being dark now, Major Murchison decided to move a short distance away to a night-time lying-up position where Marty Pelling could 'breathe a deep sigh of relief and make the decision that it was high time he took up smoking'.

South, at Garmsir, the IX Battle Group was planning a deliberate operation, designed not only to keep the Taliban at arm's length from the town but also to gain intelligence. The outline plan for *Operation Glacier 4* was to attack and destroy a major Taliban headquarters and stronghold south of the town. It would be a notable operation, for it was the first time that UK-trained Afghan artillery units were in support. Colonel Magowan's aim was to disrupt Taliban activity, deny them sanctuary and destroy the defensive positions from which they had routinely launched attacks against Afghan forces. It was hoped that large groups of the enemy would be forced far south of Garmsir to ensure that the all-important redevelopment could continue. *Operation Glacier 4* was to be part of the commando brigade's continuing offensive to disrupt Taliban command chains, lines of communications and their ability to resupply and rearm. With the first use of Afghan artillery it would also mark a significant step towards enabling full combat capability for the Afghan National Army; which would lead, ultimately, to Afghans providing their own security.

On his return to Garmsir from the Jugroom Fort raid, Colonel Robert Magowan knew the enemy were 'really pissed off', for while he and his team had been south, the Taliban had been surging into the area. Already a number of new positions had been identified via a variety of classified sources but also with the IX Group's own 'eye in the sky' – a tethered white balloon, like a small barrage balloon, called *Revivor*, that had a camera attached below it for close-protection observation.

As the IX Battle Group's TAC HQ were in the middle of a planning meeting for *Glacier 4*, and a patrol was traversing the southerly compounds in the outskirts of the town, the Taliban conducted a well-coordinated attack on JTAC Hill. The enemy hit the hill with just about everything they had from a rallying point 400 metres away from Garmsir DC and just over the river. The patrol out to the south was moving past the Taliban when it found itself split in two as the enemy attack was launched. Even the *Revivor* balloon took an RPG, which sent it flapping about the place while a number of RPGs landed in the camp. The Taliban had chosen a good spot from which to attack and their aim into the camp was good; some of the TAC HQ staff had yet to come under such sustained fire and hear for the first time the noise of the 'Afghan bees' as the bullets flew past. Three of Robert Magowan's TAC HQ were injured and had to be evacuated; but not the RSM, Colin Hearn, despite taking shrapnel in his toe. The attack was fiercely repelled, with the patrol on the ground able to call in mortar fire from JTAC Hill and the snipers on top of the high ground forcing the Taliban back.

Zulu Company's new commander, Andy Lock, described *Glacier 4* as a 'large operation to relieve the DC of pressure from the south'. After Jugroom Fort, the ex-Agricultural College had been taking many 'incomers', with the intelligence picture also suggesting that a number of senior Pakistanis had been killed and that they were now rallying more troops for a retaliatory strike. All the time Lock was receiving intelligence reports saying, for example, 'Be prepared to take a pasting.' With no CSM – Mr Shepherd was on leave – he was rotating marines into Bastion for a few days' rest, but now had to cancel those marines who were about to go and order back, in quick time, those that were already away. He had managed to get two troops out but kept the third back; 'so a bit of leadership was required as it was pretty pissing-off for the guys who were about to take their deserved period of rest.

'Every so often I would be woken in the middle of the night to be told that there were "a hundred and fifty foreign fighters in your back garden, on their way to attack you" but I never knew how many there really were out there. Whoever they were, we could always tell the difference between the Tier 2 local Taliban and the Pakistanis and Baluch – definitely. They were much better trained.'

To launch *Glacier 4* it was decided to mount a massive bombardment of the three objective areas to the south of Garmsir: Bronze, Silver and Gold; then Zulu Company would attack at night with C Squadron firing them in from across the river. India Company, brought down from Lashkar Gah for the duration, would protect the DC. The Brigade Reconnaissance Force would remain in reserve.

Colonel Magowan knew that in Afghanistan all the enemy wanted to do – were desperate to do – was to get behind any coalition force, 'to cut off and kill the infidel'. He knew, too, that as soon as his men were among the compounds and contained by the close *bocage* country, that was precisely what would happen. His plan, therefore, was to move sufficiently far in to attract Taliban attention but not so far that his men then ran the risk of being surrounded; 'If I don't go in hard enough they'll just back off but if I go in too hard they will cut me off.' He summarised the plan as: 'Go in. Attract the target. Target comes to me. I pull out and I then use B-1B bombers and Apache. As I have fantastic eyes in the sky I can destroy any target I like with great precision.'

Studying clear video footage beamed direct to his operations room in the Agricultural Centre, the colonel could devote much care to target identification. He and his team watched men 'not out walking their dogs' but Taliban moving back from positions roughly 500 metres to the south of the DC. As the enemy ambled across the pitted countryside they could be seen joining gaggles of others emerging from bunkers and tunnel entrances: all assessed as Taliban moving south 'for a bit of R-and-R'. During the night of 17/18 February the timing was deemed right. Just before H Hour a 2,000-pound bomb was dropped on the main bunker system in a determined, calculated, pre-emptive strike. The target, which had been identified using the company's ISTAR assets, was destroyed even before Zulu Company began their advance.

Zulu Company raced through to their objectives on Bronze, Silver and Gold, meeting no opposition on the way; on their arrival, the targets were empty. It appeared that the strike had been successful – 'all the enemy just died when the bomb hit' – but too successful, for there was nothing left for Zulu Company to do. Major Lock wanted to hold the position and see if any Taliban appeared, trying to fight back, as the plan had originally called for, but the colonel – listening in to the ICOM chatter – refused. He could hear the Taliban saying, 'There are too many of them. Too many of them,' so he knew that his marines should be brought in, for there was nothing further they could do. The company was withdrawn to the north of the road, then back into FOB Delhi.

Despite this, '*Glacier 4*,' Magowan believed, 'did not really work. We didn't get the timing right. Jugroom was probably too hard, too fast. *Glacier 4* not hard enough.' The solution, 'which worked out perfectly', was to be *Operation Glacier 5* conducted on 7 March.

Meanwhile, further north, on 27 February, Juliet Company was busy conducting a lengthy combat recce patrol to the east of Now Zad during one of *Alabaster*'s enduring operations. In order to relieve pressure on Lima Company at Now Zad and 'to give the Taliban something else to think about', Ewen Murchison was ordered to conduct an operation to disrupt the enemy and interdict his lines of communication. This involved 'MOGing' his company north towards Musa Qal'eh before turning westwards across wide-open country. He was heading for 'Tom', a narrow mountain pass that bisects Crocodile Ridge and a route they had previously used. As the leading troops approached the ridge from the flat desert terrain they dismounted from their Vikings to began a cautious approach on foot.

Marine 'Gav' Teece was with the point section, scouting well ahead of his Viking. Teece had had an interesting start to his career. During training at Lympstone he had fallen from the death slide and, after two years of remedial training, was medically discharged, yet his desire to be a Royal Marine was not dampened. Seeking private treatment and after a long recuperation, he rejoined and passed recruit training. On arrival at Camp Bastion he had helped in the store but, constantly stressing his well-proven desire to be a rifleman, managed to 'escape', as he put it. He joined Juliet Company in January and quickly became known as 'an excellent GPMG gunner'.

Teece, observing all the time, was about to cross a small wadi bed when, looking to his left, he spotted a rock on the edge that had been pushed beneath a small ledge – clearly by hand. As it looked out of place, he shouted to everyone to stop. Then he summoned 'the bomb bloke' who crept forward to have a closer look. Wires led from behind the rock and as the marine and the disposal expert retired to a safer distance to sum up the threat, the detonator, but not the main charge, blew. The two men were convinced that 'somebody was watching us from up the hill', the foothills to the main mountain, because as they pulled back further there was a second, massive, explosion and 'the bomb bloke' said to Teece that 'it was probably a bar mine or equivalent'.

The detonation was clearly the signal for an ambush to be launched. Before the dust had settled back into the sand and rocks, Murchison's company came under a sustained and violent attack from three separate firing points. This was a well-armed group, with not just small arms, and mortars started raining down upon the advancing British marines, but also the heavier GPMGs and the much-feared 107mm rockets causing havoc. The Taliban had chosen their ambush site well: along a small track just wide enough for one vehicle, the marines couldn't make their superior weaponry and manoeuvrability count. Determined, though, to push on to the west of Crocodile Ridge, Murchison summoned an ammunition technical officer – 'the bomb bloke' – to clamber forwards to clear the remainder of the route. Two more 107s were fired as the team securing the helicopter landing site were engaged by mortars but, their firing points identified, these were neutralised using HMG and GPMG.

Sergeant Gaz Watford, the Platoons Weapons Instructor, 'pinged' a heat source from what looked like a cave on the side of a rock face and fired a Javelin at that. Sam Vocea, a Fijian on loan from the brigade command support group as a driver, took charge of a 51mm mortar; creeping out from cover, so exposing himself to enemy fire, he popped off rounds on to a number of targets, destroying one enemy position with a 'remarkable' 800-metre shot. The marines continued the fight back, but the enemy were difficult to shift, as – this time – they had the advantage of the higher ground. It was very difficult for the marines to spot the Taliban's firing positions among the rocks and mountain crags, even though as the sun rose during the morning the light improved. Looking up the pass, all the CSM could see was a 'sea of tracer all over the place'. A second remote-controlled IED detonated a little later, followed by yet more accurate SAF, RPG and mortars that kept the troop pinned down in the mountain pass. Just after midday, and still under accurate fire, Ewen Murchison reluctantly gave the order for his the company to withdraw from 'Tom'.

This it did under its own covering fire as, for once, the Apaches could not get too close. The CSM, Marty Pelling, busy on the mortar line with resupplying ammunition throughout the long contact felt, charitably, that, due to the proximity of the high ground either side they, unusually, needed to exercise some caution.

Although they took no casualties in extracting the men from a mountain-pass ambush, the company commander was considerably frustrated: 'They guessed where I was going, but whether or not they had all three routes though the ridge covered I don't know – but we chose the wrong one.' In line with one of his less formal tasks the sergeant major was able to placate his boss by saying, 'You are joking – what we have just done here is to extract from an ambush with no dramas. It was an awesome thing to do.'

The lack of casualties was due to good fieldcraft, for almost every vehicle had been struck by rounds which had also passed through all the bits and pieces strapped to them, such as jerry cans. Sam Vocea, later checking his kit, found his bivouac and sleeping bag had a hole burned clean through; thinking it the work of a 'clumsy smoker' he was about to go on the warpath, when he discovered that an enemy 7.62 tracer round had melted into the fabric. He now wears it with his dog tags.

The following day the company returned to carry out a cordon-and-search operation on the nearby compounds and derelict buildings. In one they found 10,000 rounds and several Chinese 107mm rockets. None of the locals working in the fields nearby knew anything about them, of course, so the company's Afghan interpreter suggested that the British should take two of their cattle – or two of the villagers – and shoot them to get some answers. 'We're not the Russians,' replied Murchison, 'and we don't execute people.'

As Juliet Company pushed north, the rolling scrub desert gradually gave way to the mountains that severely constrained its freedom of movement: it was far enough. 'The important thing about MOGs,' Murchison felt, 'was not knowing much about the areas we were going to. We were feeling our way and had a 50/50 chance when pitching up of being offered a cup of tea or being in a threeway ambush. We were always configured for the worst case.'

Juliet Company made this north-west corner of Helmand its own stamping-ground, for other MOGs – combat recce patrols – were conducted during a series of ten-day excursions into this harsh and unforgiving Taliban territory. By travelling further north up the Baghran valley than any coalition forces had done so far, they were a serious thorn in the Taliban's side; this area was, or had been, theirs alone. For instance, Murchison received a report that Kisney housed a possible enemy R-and-R and logistic site, and so it was here that he held a 'brilliant' *shura* attended by about fifty men, who were all probably Taliban 'as there was not a woman or child in sight'.

'I thought it was more of a training camp than a civilian village and they let me in partly because when I arrived they were all in the mosque – probably having a lesson. I sat down with my helmet off 100 kilometres away from the nearest coalition base and explained that we were neither Russian nor American and that we were not there to eradicate the poppy. Only to provide security. Not to build wells either, but if they told me what they wanted and they complied with us they might get some construction and redevelopment.

'What are not seen in the photograph are the Vikings and hordes of marines behind me, so there was little chance of me being taken hostage; my men would not have let me go alone any distance at all.

There was a Canadian who went to such a meeting and he was chopped in the back of the head with an axe. Whether true or not, it focused the mind when you sat down with these people. I always had a bodyguard right behind me watching every movement.'

The men told the major that there were, of course, no Taliban in the area, and that they wanted the British to go away, for two women and children had been killed in crossfire the previous year; they just wanted to be left alone. But the atmosphere was more wary than tense, and although neither side was able – or willing – to offer the other anything, the *shura* ended on a friendly note.

During their return the marines travelled past Shurakay, on the edge of the green zone, which the Taliban were using as a base for strikes against Sangin and from where the Taliban now launched an ambush. Up to thirty of them opened up at close range with the marines rapidly de-bussing and returning fire. While a flanking section set off to steal round the side of the sand dunes where the Taliban had positioned themselves, the WMIKs hurled fire in their direction. The pressure worked and the Taliban retreated into the empty village; the marines called in artillery fire to target them but despite the majority of the Taliban having been killed in the fire-fight they continued to mortar and RPG the company. If nothing else, they were tenacious in defeat.

On one of these combat patrols – a few kilometres north of Now Zad – Marine Andy Thomas was run over by a Viking as Juliet Company came into a 'fairly heavy contact' in open ploughed fields. The feared 'Man down' call was passed across the company radio net: the CSM's cue to make his way to 2 Troop's Viking. There he found that both cabs – twelve tonnes in all – had run over the marine as he took up a firing position to the rear: the driver, seeing his other 'passengers' de-buss and move off to the right, assumed that all was clear behind.

Nobody could gauge how badly Thomas was hurt, although it was visibly obvious that the side of his body had been deformed.

He had undoubtedly been 'squashed in and his eyes were glazed over'. There was much blood and mucous around his mouth, he could not see and he was barely conscious. 'Thomas was a very frightened man and occasionally just conscious enough to know he was badly hurt.' The main concern – thankfully untrue – was that he had severe internal haemorrhaging and was bleeding to death.

From 'point of wound' to 'wheels up in the Chinook' took only forty-five minutes; an elemental fact that allowed him to make a full recovery. CSM Pelling, who had orchestrated his recovery from the battlefield, knew only at the time that he had a broken collar bone. He was to make a full recovery before attending a driver's course 'so he can get his own back!'

7
MARCH: RETAINING THE INITIATIVE

'This was, without doubt, the bloodiest and hardest time we had.'

If February had been a month of musical chairs and exploitation, then March – the penultimate month of the commando brigade's deployment – was a period of even harder fighting and a month of tragedies, with four men killed in action.

Once *Operation Kryptonite* had been successfully completed and with Kilo Company settled into FOB Zeebrugge and its associated observation posts on Athens, Normandy and Sparrowhawk, Mike Company flew to Camp Bastion for two days' respite that included a 'company meal' together and a can of beer. The CO encouraged his men to take a mental break before their next mission. Rumours – based on intercepted ICOM chatter – were beginning to suggest that the Sangin ceasefire might collapse, with Lieutenant Colonel Matt Holmes knowing that if the 'fragile peace' did end he would need battle-hardened men to counter the expected contacts. Mike Company was 'warned off' to relieve Charlie Company, The Rifles – whose successful but quiet stint was coming to an end – which they did during *Operation Platinum 2* on 24 February.

Arriving in the large Sangin District Centre the Fire Support Group, under the command of Sergeant Jay Layton, took over the .5-calibre machine guns in their firing points. As a precaution

Corporal Mackinley, a heavy-weapons specialist, stripped, cleaned and oiled them before announcing that all weapons were ready to face any onslaught.

Immediately the company began to build up the 'panoramics', that is, getting to know the ground and the situation, aided by a most useful handover from Charlie Company's commander, who passed on not only what he had discussed with the elders but also his perceptive interpretations. Other good sources in the area relayed much of what else was happening beyond the District Centre's walls. ICOM chatter suggested that the Taliban were, amazingly, tunnelling beneath those very walls in a desire to invade and capture the DC itself, but this frightening prospect was countered by the marines' keenness to engage in a close-quarter battle. Regular listening to the chatter also indicated that there were over 400 fighters in Sangin – an area they used as a base station before moving off to Kajaki and elsewhere, but the Taliban were well aware that their communications had been compromised, so they might have been bluffing when talking of such large numbers.

The ceasefire broke, as the rumours had suggested it would, on Mike Company's second day in Sangin, with a well-coordinated and accurate mortar attack. This first of *hundreds* of attacks came from the north and south with mortar bombs landing inside and outside the compound, mixed in with RPGs that were fired from close by. The main effort, as the marines expected it would be, was against the FSG tower, 'because that had the most firepower and was the biggest building, with the Taliban knowing that during any contact it would be manned'. Obviously they believed that if they could destroy the tower, their chances of capturing the DC would increase. The Paras, when they had been based there, had painted a sign at the foot of the stairs leading to the roof and then tower, which read, 'Thirteen Steps to Heaven – Thirteen Steps to Hell'.

On that first day of fighting the marines had considerable difficulty in identifying the enemy's positions, for many were at a great distance and it was, as yet, unfamiliar territory. Over the next fifteen days the attacks were to increase massively, both in intensity and duration. An addition to the Taliban's already impressive arsenal that the marines came to dread was the AGS 17 grenade launcher with its 1,700-metre range: these vicious 30mm grenades were fired across the tower and into the top-floor sandbagged bunkers, where they exploded into steel shards. It was not long before these missiles began to take their toll.

On the fourth day the DC stood-to for yet another mortar attack with Jay Layton and his marines in their positions on the roof. His FSG team in the tower included 'two lads from 29 Commando Regiment Royal Artillery', with whom he had worked closely at Kajaki. The mortars started thudding into the ground inside the compound while the men watched closely from their bunkers, for, as Sergeant Layton reminded them, the RPG men would have to be in close and make themselves visible if they were to shoot at the tower. 'Keep watching your arcs of fire,' he shouted – then an 81mm mortar bomb hit the tower. Layton was standing on the far side of the structure, but Nick Gilbert was only five metres away when it exploded. The high-explosive round shattered, sending thousands of pieces of hot metal into the area around it, pinging off the metal fittings and thudding into sandbags – and flesh. Gilbert took masses of fragmentation up his legs and partially in his back as he was flung sideways by the force of the explosion. One of the lads shouted for the medic, who came sprinting round the roof, while Jay Layton urged the men to continue spotting the enemy, to keep watching. Gilbert was lying face-down by a bunker with blood on the ground around him, but he was still conscious. FFDs, first field dressings, were clamped over the wounds before he was carried, as best they could, to the bottom of the steps. Meanwhile the Taliban appeared closer, readying their RPGs, allowing the marines to fire back at them at last.

Fortunately Gilbert wasn't too serious a casualty. He was in a stable condition, but he did need proper hospital treatment. Unfortunately he could not be flown out that day as the risk was too great; the Taliban, desperate to shoot down a helicopter, had come in close, probably with that aim in mind. A Chinook could not land under those conditions – the DC was surrounded – but that night a huge weight of 105mm fire was laid down against all the major targets identified during the day. Air support was also available in the dark. The Chinook arrived the next morning to be greeted by a few RPG rounds, but they all missed and the crew managed to lift Gilbert away to Camp Bastion for treatment.

The general situation continued as bad, with junior commanders trying to stand-down their men as often as they could; but with contacts throughout each day – and often well into the night – most time was spent under fire. Jay Layton, responsible for the exposed positions, tried to work a rotation between his sentries and lookouts but it was hard going. With four or five marines in the tower at any one moment he noted that there were certain vulnerable times during the day when the Taliban would try to ensure they would definitely be hit. 'When each contact kicked off we had to climb the stairs with the guys bloody terrified. They were quite right, for it took a lot of courage to run up and across the roof with all that lead and steel flying across the top of it.'

On 3 March Layton was on the roof with four lads during one of the more vulnerable times of day – about noon or a bit later – moving around, spending half an hour with each sentry in his position. As he watched towards the south before looking back at the river he heard a crack and, to his left, caught a brief flash 'coming from a little spot we called Heart-shaped Tree'. 'Take cover – take cover,' he shouted before throwing himself to the floor as a huge explosion sent 'bits of steel screaming and whirling, flying all over the place'. Crawling out from behind a sandbag he stood amid the settling dust and called out, 'Is everyone OK? Everyone OK?'

Everyone was not 'OK', for 'Paddy' McLaughlin – a lance bombardier with 29 Commando – was lying on the roof's floor. He had taken a direct impact from an RPG and been killed instantly. Then Ross Clark – another lance bombardier with 29 Commando – came staggering towards Layton, muttering, 'Jay, I think I've been hit.'

'I grabbed hold of him and pulled him into cover behind some sandbags before turning him around and asking, "Where have you been hit?" but he didn't answer. When I pulled his body armour aside it was obvious. He had a large wound beneath his armpit, going in towards his heart. I tried to give him first aid – morphine, FFDs – and kept on talking to him, desperate to keep him breathing, as he was becoming very laboured. Although there was still one hell of a lot of small arms going past I shouted for the medics. But I knew he was unlikely to live. He lay in my arms while his breaths became weaker and weaker as he drifted in and out of consciousness. The medics came running up the stairs looking for Paddy, not knowing that we had two casualties on the tower. I think Ross was a dead man but just hoped the doc could do something with him – you just hope for that. Don't you?'

With the medics doing their best for Ross, Layton knew he was only in the way so, as he had other duties to perform, ran across the rooftop to encourage his team to continue engaging any target they could see: not that they needed any further spur to their efforts. Then, as the firefight petered out he was able to tell his men what had happened. In such a tight community the loss of the two popular gunners was keenly felt.

Although the attacks continued to take place mostly in daylight, the Taliban used the dark to creep in towards the DC, apparently unaware that every move was being watched and recorded. The Javelin's command launch unit possesses an excellent night sight through which Mike Company's marines could 'see everything that was going on around us. Thanks to that we were ambushing them at night with some brilliant results.' As well as launching attacks from the DC, the marines called in air and artillery support to batter the Taliban's positions.

The ICOM chatter, especially towards the end of the first fifteen days, suggested that the Taliban were suffering badly, 'which perhaps was unsurprising when you consider the amount of munitions we were expending on them'. But their tenacity and willingness to fight remained strong. The marines were aware of that, kept their guard and even their sense of humour, despite the suddenness of each attack, the intensity of the 'incomers' and the consequently difficult living-conditions.

The attrition continued when, a day or so later, two more men were wounded on the rooftop: Corporal Phil Mills and Marine Craig Watson were both hit by enemy grenades. Watson was the marginally less 'lucky' of the two for he was standing by his own grenade machine gun when it was hit. The explosion caught him down his left side and left shoulder, blowing him into a corner of sandbags. Saved by his body armour, another incoming grenade then exploded, causing more blast wounds and severing his Achilles tendon.

Sergeant Layton was again on hand, furious that once more his 'lads' were taking these blows. Even as the medics came rushing up the 'stairway to hell' to answer the call, he was determinedly scouring the area from where the missile had been fired, for he wanted more than anything to 'neutralise' the grenade gun. After the losses of Ross and Paddy he had asked for two snipers to assist him on the roof; 'they were good, for they could concentrate on greater distances and began dropping Taliban on a regular basis: especially at Heart-shaped Tree where the enemy thought we couldn't see him. Charlie, a sniper, managed to spot him. Took the guy out straight away with his .338 at about 210 metres. Close in and not that difficult – but useful.'

Needing urgent treatment, the two casualties were lifted out that day with the Chinook pilot flying in under fire with, as always, Apaches above providing 'top cover'.

Following the loss of two of his lance bombardiers, Lieutenant Colonel Neil Wilson, the CO of 29 Commando Regiment, sent in a new team headed by Sergeant Major Mick Smith, well known to Mike Company from the Kajaki days and regarded as 'a brilliant bloke'. Following the briefing that Sergeant Layton gave him, Smith was under no illusions about the position he was inheriting.

On 8 March Mick Smith had made some pitta bread using flour he had brought with him and offered Jay Layton a 'slice' for breakfast. He had just enough time to savour this unexpected and welcome luxury before 'running off to see the OC for morning orders'. When he reached the bottom of the steps he removed his body armour, as it was a warm day and up till then reasonably quiet. Having received his orders – and a mug of tea – he headed back up the tower, donning his armour as he went up.

Almost as soon as he reached the top the incoming fire started. This time a grenade exploded about three feet from Jay Layton as he was observing incoming enemy rounds: luckily the blast 'went the other way' and he was able to dive into a bunker next to Marine Dave Pennington, where they both lay with their hands over their heads as more grenades exploded around them, swiftly degrading the sandbag protection. It was only a matter of time: a grenade entered the open door and exploded inside the bunker, although luckily the force of the blast – apart from the noise, heightened by the small space in which it exploded – was absorbed by the sandbags. With his brain temporarily confused and his ears singing, the only other sounds to reach Layton were the 'Scouse' tones of Sergeant Mick Smith's loud voice.

'What the fucking hell are you doing in there? Get up. Come on, you poofs, start looking around.'

'What am I doing in here? I nearly got killed, you silly bastard!'

Mick laughed, a broad smile on his big face, and said, 'Come on, let's engage the bastards.' He walked out from the bunker, seemingly unconcerned by the bullets flying about, to start calling in artillery shells and air bursts as close as he could to flatten the enemy.

Things calmed down for a while now that the guns were keeping the Taliban at bay, so once more Layton stepped down and removed his kit. Suddenly 'there was a fucking great explosion on the roof and everyone shouting, "Get on the rooftop. On the rooftop," so I donned my kit and as I reached the base of the stairs I saw someone lying at the top where the rounds were zipping through in a massive wave of small-arms tracer. You have to be careful because the second floor is open. Someone shouted, "Don't go up the stairs. The rounds are coming in at the top." After a pause I shouted, "Right, follow me," and ran up to find that it was Mick lying on the floor. He had been hit in the face by a grenade.'

Bombardier 'Brum' Jennings was trying to give Smith first aid – he was alive, just, but in an horrific state. Layton made sure the lads stayed with him, shouted for a medic, then ran round the top of the roof as another explosion hit the tower followed by yet a further shout for a medic. 'One more of my guys had been hit by an RPG. Marine Giovanni Brice was lying on his stomach with very bad shrapnel up his lower back and down his legs. Seemed OK otherwise, so I got a first-aid guy to deal with him.

'We carried on with that contact – managed to get them casevacced – while we were still up on the tower trying to engage targets and getting aircraft to drop 1,000-pounders within 200 metres of the DC. It just went on that day. Non-stop.'

Mick Smith later died of his injuries in hospital at Camp Bastion.

Bombardier 'Brum' Jennings, having already lost two of his 'oppos' and who was to win the MC, was 'an awesome guy' whose task was to call in aircraft. All Jay Layton – also awarded the MC – had to say was, 'I want that hit and that hit and Brum would bring in the right aircraft for the job.' Earlier in the tour, when in a convoy, his best friend had been killed in a landmine explosion, 'so he had been through the mill more than most'.

The Royal Engineers of 59 Squadron RE were as admired as their gunner counterparts. 'They would shout up to the tower to see if we wanted any battle casualty replacements, and if they weren't mending the sandbagged positions under fire they were taking their place in the firing positions.' Sergeant Layton summed up: 'It was a whole team effort because you seemed to be in your own little world on that rooftop. It's not good losing a quarter of your guys there from enemy fire. Not good at all. But no one let us down. All the boys were fantastic. Very, very brave young men.'

Two days after Mick Smith was killed, the ICOM chatter began suggesting that the Taliban were interested in calling a second 'truce'. As the tempo of attacks dropped off, the marines of Mike Company believed this was due to the enemy running out of ammunition and men. They had certainly taken 'one hell of a hammering' and 'were always keen to wave the white flag if they needed a chance to resupply'. The ICOM chatter – although open to bluff and double bluff – was useful. More than once, while under an artillery barrage, the Taliban would scream that they were being shelled. The order issued out to the gun line immediately would be, 'Fire another five rounds,' and then the marines would watch to see them fall.

In contrast to the Taliban, the besieged Mike Company found that although their fifteen days in Sangin were difficult, they had all they needed. 'We had enough ammunition except grenades for the grenade machine gun but, as it was a new piece of equipment, that was hardly surprising. The WMIKs were good and maybe we could have done with some more of those. Especially the newer E-WMIKs with the V-shaped hulls that saved a few lads' lives. Over a landmine the wheels would be blown off instead of the whole engine block killing everyone.'

Jay Layton's opinion on the Viking was not unusual: 'I preferred the WMIK but the Vikings certainly saved a lot of people's lives.' Overall, his view was that 'all I ever felt we wanted was more WMIKs, more aircraft, and more legs on the ground'.

On 8 March a much-exhausted Mike Company was relieved by Charlie Company of the Royal Regiment of Fusiliers.

While Mike Company were enduring their fifteen difficult days in Sangin, Neil Sutherland's Kilo Company at Kajaki were about to take the war to the enemy during *Operation Knight*: a deliberate company group assault.

Lieutenant Colonel Matt Holmes felt that without sufficient aviation lift he could not reinforce Kajaki any more. A series of highly kinetic operations had certainly reduced direct Taliban influence over the dam but, further out, their continued presence remained a running sore by preventing the civilian population from returning to their homes and fields. It was now up to the resident company.

Neil Sutherland realised that the hardest task his company would face throughout the deployment would be *Operation Knight*, aimed at clearing the haphazard maze of some sixty compounds that made up the village of Bagar Kheyl from where, during *Operation Kryptonite*, they had taken significant fire. Although it was believed that some civilians remained there; so, once his men had broken in, a number of options would present themselves. Either they were going to go through in a 'fighting in a built-up area' -type operation or, if they encountered civilians, they would try and establish a *shura* with the elders. His outline plan was to clear Bagar Kheyl with a view to 'not only engaging the Taliban and disrupting their forward lines but to also engage with any locals they could find'. He wanted to impress on them why his company was operating as it did. In the end Sutherland was able to confirm that was indeed 'the most difficult thing we ever undertook'.

As always Sutherland did not have the combat power to stay *in situ* to establish a patrol base which would have allowed him to establish his credentials with any civilians before he was obliged to leave. On the other hand, if he took on the Taliban, then he would have to fight them in the hope that the civilians would push off quickly and in advance. The marines were used to this, regarding such movements as 'always a combat indicator anyway'.

'If it was the Taliban, then we were going to clear them out. Identify all their firing points, take GPS positions of every one and next time we take fire, drop bombs directly on to those positions from the B-1B bombers. As we wanted to significantly disrupt the Taliban in this area we would have to set up sniper ambushes and pop up here and there. Keep them guessing and prevent them from infiltrating back behind us.'

In the hours before first light on 6 March Kilo Company moved out to establish a gun line on Shrine Hill while the fire-support WMIKs pushed forwards to the south of Magar Kheyl across the now dried-out wadi beds. This gun line's duty was to isolate Makar Kheyl and prevent any movement into the target area, about one kilometre away. The observation posts above Zeebrugge were providing the second layer of defence, effectively cutting off Bagar Kheyl entirely from its surroundings.

Sutherland now ordered 4 and 5 Troops to the south of Bagar Kheyl with his Tactical Headquarters (TAC) and an ANA troop behind: overhead he had two F-15 fighters and two Apaches. Immediately prior to first light the Apaches were to 'come in very quickly and have a look at the first two compounds, which were the break-in compounds'. With no sign of life detected either by the helicopters or the OPs to the south, the marines began their move across the start line at 0600 – moments before the sun started to rise. The plan was for Lieutenant Chris Payne's 4 Troop to break into the compounds to the west, simultaneously with Lieutenant Luke Kenny's 5 Troop breaking into the eastern compounds. The ground leading up to the break-in compounds was open and exposed. The compound to the west was on high ground with purpose-built firing holes spread along the tall walls; to the east of the village were large open fields that sloped gradually away down-hill with a further uncleared village to the north.

The company commander needed this simultaneous break-in for, despite his precautions, he was still expecting significant fire from Magar Kheyl. If a 'second' troop had been caught in the open ground after the fighting started they would have a serious prob-lem, for there was little cover. The added difficulty was that as no movement had been detected, no pre-emptive fire, paving the way for the assault, was possible. Sutherland had, however, deployed his Desert Hawk UAV over the compounds, so he and his troop commanders knew precisely where to break into a compound; on the supposition that there might be civilians he did not want his men blowing their way into their living spaces.

The company crossed the start line with two troops up and Coy TAC back. Both troops, each having dropped off a section in reserve, now moved simultaneously towards their individual objec-tives. Leading 5 Troop was Corporal Al Hewett's seven-strong section while his friend, Corporal Si Willey, was commanding 4 Troop's leading section.

The previous day Hewett had monitored the high-resolution camera screen and identified where, on the ground, he could place a covering group of general-purpose and light machine guns and where he could best place his assaulting pairs of marines. He was again down to seven men thanks to R-and-R and, with no reinforce-ments, he needed to be sure where each man would be. Hewett and Willey knew that movement would be hard and exposed.

Hewett's ordered approach took his section north-west from Zeebrugge across a large, dry wadi – the M1 – before bypassing various hamlets that had been cleared during earlier operations. He was aiming for a holding area in dead ground approximately 500 metres from the objective. Here 5 Troop's reserve, 3 Section under Corporal Den Dennis, moved into a nearby compound with the troop sergeant and the 51mm mortar. Corporal Willey's section continued to move towards its own line of departure that faced their break-in compound across the same 150 metres of open and exposed ground. Moving twenty metres behind each leading section were teams of Royal Engineers carrying mouse-hole charges. As the manoeuvre-and-fire boundary between the two troops, Sutherland had chosen a small road which split the village in two.

When all was in place, the order was given to fix bayonets and advance: within fifty metres 'the world erupted' as the Taliban launched every weapon at their disposal against their foe. The barrage was so effective that even the Company TAC HQ came under fire.

Despite the fact that both leading sections were taking significant fire that pinned down the complete company for forty minutes, 4 Troop's leading section continued its advance. The devastating small-arms onslaught, though, drew the first casualties. Twenty-two-year-old Marine Ben Reddy was killed instantly, while Lance Corporal High, the section second-in-command, was hit by three 7.62 rounds in his chest armour. He would have been unscathed but one round ricocheted through to his thigh and he fell. Much of the fire into Willey's section was coming from 5 Troop's objective. Al Hewett's section had been targeted from a village to the north with one rocket falling between him and Corporal Webster. As they had covered only one-third of the open approach – and were now under a maelstrom of small-arms and rocket fire – Hewett was ordered by his troop commander to peel back to the dead ground he had just left; a manoeuvre successfully covered by Corporal Webster's section which itself was also 'pepper-potting' into better fire positions. Over to the west, across the troop boundary, 4 Troop's reduced point section reached the first compound wall and were desperately attempting their break-in using bar mines, but the Taliban had reinforced their defences and the powerful explosives were making little impression. All the while the enemy continued to fire at them from 5 Troop's objectives.

With 5 Troop out of the way, Major Sutherland, commanding from 100 metres back on the edge of the wadi, ordered the Apaches forward. The helicopters duly swung in and slammed Hellfire missiles and 30mm cannon into the right-hand, northern, compound. With this enemy position now quiet, 4 Troop's leading section were able to place a third bar mine on to the left-hand compound's outer wall. At last a mouse hole was blown, enabling Willey's section to gain that first, vital foothold. Concurrently, Willey's two casualties were being picked up on stretchers and carried back to CSM Clark, who drove them in a trailer attached to his quad bike down the wadi. From there a Pinzgauer motored them as fast as it could back to the FOB and the wait for a Chinook.

The company's fire-support WMIKs were now under mortar attack from a position that had not been seen but this, too, was neutralised by the 81mm mortars.

Having achieved their break-in, 4 Troop were able to clear the next four compounds, smashing through each wall and using grenades to force out any Taliban hiding inside the near-derelict buildings. Major Sutherland moved forward with his TAC HQ to gain a clearer picture of the ground ahead and which compounds needed 'sorting'; 5 Troop were despatched to join their original route, now that the Apaches had been thoroughly 'hosing the place down' with cannon and Hellfire missiles. Meanwhile the CSM, taking charge of resupply, delivered ammo to 4 Troop. 'As we were receiving a great deal of fire from other positions, the gloves were off and we began to really fight through with fixed bayonets and grenades – at close quarters and hand to hand.'

Al Hewett's section were about to get an unwelcome taste of that hand-to-hand fighting. On his return to the dead ground, Hewett had crawled across to his troop commander and signaller to ask for a Sitrep from TAC HQ. As it was received, Luke Kenny looked at his corporal and said, 'You are going to have to take that compound.' Hewett's decision now was to leave the engineers behind for the moment, as the walls were proving too tough an obstacle this time, while they tried to break into the compound through a door or window, using grenades to clear their path.

Gathering his section around him, they zig-zagged their way forward, keeping as low as possible while dropping to fire as they covered each other in the approach to the compound walls. Behind them the reassuring *chunk-chunk* of the mortars continued as they fired ahead, to where the Taliban were judged to be, while all around them bullets buzzed past. Using this 'fire-and-manoeuvre' strategy, they successfully made their way across the killing-ground to reach shelter beneath the nearest mud and rubble wall, while the section commander concentrated all his fire 'on the three or four firing holes that the enemy were targeting us through'. The wall was no more than twenty or thirty metres in length, with the entrance to their left and an extended right-angle wall to their right forming an 'L' shape. Pressed up in a pack against the compound wall, the men all quickly prepped their gear, changing magazines and readying grenades as they moved into position for an explosive entry.

Steve Freer gripped a grenade and prepared to chuck it over while Tommo Thomas covered him. With his back to Thomas, Hewett glanced at Dan Ravenscroft and Ellis Alaimo next to him. Beyond them, next to Gav Wignal and Ducky Mallard, Hewett noticed a purpose-built firing hole cut into the concrete; he immediately told them to stick their weapons through and prepare to 'open up'. But just as he was about to give the order to 'go', a machine gun popped out of a firing point on the right-angled wall some five metres to the right. A storm of 7.62 from an AK-47 lashed out in their direction.

'As the rounds bounced off the walls towards us I heard a shout from Dan as he hit the ground and then from Tommo as he did the same. Both had been shot. Ellis and I turned and fired through the wall just as I gave the order to break in to Steve. The round that hit Dan had passed through his femur, snapping it in two. Tommo was shot in the upper arm. Steve was trying to kick the door in while I sent a "Man down" over the radio.'

Hewett had more problems on his hands, for the door was heavily barricaded and unlikely to yield without explosive charges. This left him in a dangerously exposed position: two serious casualties, a heavily defended wall with firing slits to his front and a 150-metre killing ground to his rear. With little time Hewett made the decision to move across the fire-and-movement boundary to link up with Corporal Willey. This was contrary to the company commander's orders, but in Hewett's view moving his casualties, under fire, twenty metres and out of harm's way rather than 150 metres across an exposed area raked by small arms and rockets, was the better decision. Then, with his casualties safely out of the way, an Apache swooped, firing on anything it could see moving.

Once linked up with Willey's section, Steve Freer and Ellis Alaimo were able to drag their wounded into some form of cover where, under 4 Troop's temporary protection, they stabilised Dan Ravenscroft's serious injury. Meanwhile Tommo Thomas, despite being shot in the upper arm, was determined to carry on with the assault and began firing on to the original break-in compound, helped by two Royal Engineers from 4 Troop.

With Corporal Willey's compound cleared, the marines at last had the necessary foothold into the village. Now into the compound, the men were moving forward cautiously when suddenly two Taliban emerged from a trench with overhead protection – corrugated iron covered with layers of soil – pointing an RPK machine gun at the barricaded door that Corporal Hewett's men had tried to kick or grenade their way through. Perhaps they hadn't expected the British to appear from a different spot, as they fell under a hail of bullets. The outer walls on the compounds were thicker than normal and in some places had been 'doubled' to create a killing-channel. 'If,' as Hewett discovered, 'we had blown a charge through the initial wall, the enemy would have been able to catch us in between that and the next one, still trying to get through.'

The men rapidly spread though the maze, checking behind every wall in case there were other surprises waiting for them. Down the side of a wall they came across the body of the man who had shot Dan Ravenscroft and Tommo, killed by the attack helicopter as he tried to flee further into the village.

Deeper into the compounds the extent of the defences were horrifyingly clear. The tunnel system ran the length of the village; some tunnels had steps leading down into them, while others were a combination of covered-in drainage ditches and sewers. Some tunnels came up into rooms with overhead cover.

An exhausting and attritional battle followed, with both sides trying to gain an upper hand as they fought though and around the compounds. The Taliban had the advantage of knowing the ground and of being able to use the remaining defence systems; the British had the advantage of superior training and stamina. Al Hewett's section found 'holes in the walls from where they were channelling the lads by fire as they came in. Their determination probably surprised me most of all. We had to clear every compound thoroughly; for example the MFC and TAC were moving in a compound which they thought had been cleared when a Taliban appeared from a tunnel about ten metres away. The MFC shot him before he could shoot first. Another Taliban was hiding in a clay oven and a lad saw the movement and got him too.'

As the long day wore on, a seamless and natural interchanging of jobs became a common theme, with individuals being attached to different sections and engineers: some switched from providing offensive support, to clearing compounds and firing points. This ability to chop and change as individual battles – and casualties – dictated was only possible through having confidence in their own skills; it also emphasised the importance of standard operating procedures based on the realities of battle and not dogma.

The Apaches were not able to fire during much of this combat, as the marines were too close for this to be safe; but they could continue to hover overhead and guide the men around obstacles that they could spot. The Taliban seemed unaware of this possibility; Hewett was contacted by a pilot who had observed Taliban in one of the alleyways, clutching two PKMs and ready to launch an ambush as the British were about to pass. Corporal Willey shinned onto a roof, moved across and dropped a couple of grenades over the edge into the alleyway; 'Job done'.

In the village the fighting was 'bloody' as the marines roamed through the dwellings. Neil Sutherland now brought the ANA troop up to picket the buildings and tunnels his men had cleared while relaying constant Sitreps back to his second-in-command in Zeebrugge. Then, well into further compounds, the company commander was able to establish a portable satellite link direct to Colonel Holmes. The two discussed reinforcement options to exploit the success: Cdo TAC HQ with Ops One Company or the Helmand Reserve Troop. Given the time it would take to deploy Cdo TAC HQ and a full company to Kajaki, and to maintain the momentum, Sutherland asked for 'a troop – now!' and shortly afterwards a platoon from The Rifles landed by Shrine Hill. Escorted forward by the CSM, Nobby Clark, they were used as additional pickets in support of Kilo Company, who had fought through and secured nearly fifty compounds in some of the fiercest fighting 42 Commando had experienced.

Even after the break-in battle, when the company came under effective and withering enemy fire, numerous Taliban had remained in their well-prepared positions with some re-emerging from tunnels inside the compounds to continue the fight. The fighting in the compounds was at close quarters and limited only by the length of time the Apaches could loiter. Yet in that time Kilo Company had achieved much but now reports that Taliban reinforcements were approaching from the north were becoming an added concern.

At about 1400 the company conducted a controlled extraction taking GPS grid positions of all known Taliban firing and sentry positions – 'for future targeting should it be necessary'. Once the men had completed their withdrawal F-15 aircraft dropped several 1,000- and 2,000-pound bombs, collapsing those tunnel systems that had been found.

Neil Sutherland had every reason to be well pleased with his men's work that day. 'I think 4 Troop used about a hundred grenades; the FST, at the edge of Bagar Kheyl to prevent the enemy being reinforced or any leakers escaping, did an excellent job.' He also praised his mortar gunners: 'We used many 51mm mortars, but the ammunition was running short, which is a pity as, although the 81mm was good and accurate, it took time to get rounds down – not much but sometimes a significant pause; whereas with the 51 you just lob the rounds down the range as soon as you like. Some of my marines were very experienced and extremely accurate. It was a junior leaders' battle: for instance Corporal Simon Willey, a 4 Troop section commander, was responsible for many of the initial break-ins, while Marine Fisher, also from 4 Troop, took over Lance Corporal High's section after he had been shot.

'There were about forty or forty-five Taliban dead by the time we had finished but, of more interest, there were two elderly civilian gentlemen who were not killed. That was good, sensible control and restraint by the young marines. As we withdrew we escorted these two gentlemen back and dropped them off by Shrine Hill; which probably wasn't where they wanted to go but they understood the reason when we said that we were going to drop bombs on the tunnel systems.

'This was, without doubt, the bloodiest and hardest time we had, definitely the hardest fight the company had and probably the hardest fight that 42 Commando had. Possibly the hardest the brigade had, but there were dozens and dozens of other little fights which in their way were just as intense, although perhaps not so intense for quite such a long time. The mere fact that it was a deliberate attack made it important, especially as the Taliban now stayed away from Bagar Kheyl. When they did try to move back we had the grids of everything and knew exactly where they were and so we just bombed them.

'We had not taken any fire from Magar Kheyl, so either they were not there or they really did not want to get involved. They could guess only too well what we would do if they had fired, for they must have been watching our progress through Bagar Kheyl.

'The Taliban were quite happy to take us on at company level and below, but towards the end of the deployment there was very little Taliban activity. It was as though they were quite happy to stow their weapons and fight another day. I think perhaps that was the case with these guys in Magar Kheyl. Of course they will fight if they are cornered and the guys in Bagar Kheyl could have moved out to the north, but they knew the Apaches would get them. The Apaches would chase them for as far as it takes. After that we kept the pressure on with sniper ambushes and clearances up to the north-east; all the time forcing them back beyond their forward lines and away from the Kajaki Dam.'

The very reason the Royal Marines were in the area in the first place.

Ben Reddy's body was flown home to a funeral service held at Windsor Castle where his father was a gardener. Prince Philip, the Captain General of the Royal Marines, attended in support of Marine Reddy's family.

The day after Kilo Company's testing work, 7 March, *Operation Glacier 5* was launched to the east of Garmsir with a similar aim: to push the Taliban away from the centres of civilian population in order for development and reconstruction to take place unmolested.

Lieutenant Colonel Rob Magowan's IX Battle Group's objective was a compound to the east of the canal, nicknamed 'Vodka'. Those enemy that had survived the B-1B's bombs or who had been 'away on R-and-R', had now moved from their hides in Bronze, Silver and Gold, crossed the canal further south and were now occupying Vodka, a kilometre-square village from which civilians had long gone. They were also in a smaller hamlet nicknamed 'Strongbow'. With no bridges in the immediate area, the Taliban used Strongbow and Vodka's compounds as firing points into the Royal Marines' left flank, believing themselves to be immune from any direct retaliation across the 'impassable' waterway. Once more the enemy misunderstood the meaning of the word 'marine'.

With so many small canals and irrigation ditches 'hatching' the area, Andy Lock's marines had conducted one crossing using rafts made out of jerrycans and wood, and had negotiated many others with similar ingenuity. These had worked, if not always in the practical sense but psychologically, for they gave notice to the Taliban that there was no obstacle too difficult for the marines to cross. Garmsir's easterly canal was, though, 20 metres wide and would need rather more elaborate arrangements.

Certainly no vehicle could cross in the immediate area, although there was a Russian-built bridge about two kilometres to the north controlled by a police checkpoint; but this was too obvious an approach. In January, by good fortune, Lock had discovered two infantry assault bridges at Camp Bastion: both were delivered although, after the operation, only one was returned. 'There were a number of ditches,' Lock explained, 'where I needed two sections. Each section is about six foot long and I often used them for the most basic of manoeuvres, but for *Glacier 5* we used seven sections. It's a very light affair with side panels, which two men can carry. They have a roller with a float and as you push it out into the stream you bolt the next section on and continue pushing until it gets to the far bank, when two very brave engineers run across and secure it.'

Andy Lock was realistic not only about the bravery of the Royal Engineers but also about the difficulties his men would face; but the task was one the whole company relished because it was one they had devised themselves.

'For my company group to negotiate an obstacle of that width and depth at night as a fighting formation and with the likely "incoming" fire was going to be tricky, although I had all sorts of wonderful suggestions from my marines about swimming across. This assault was the culmination of my time in command of Zulu because we planned it pretty much from scratch: we had also developed the target ourselves. The battle group added weight to my intelligence as plans gradually developed, but the attack was ours from start to finish as a company group. To take 150 guys across the bridge at night and then conduct an offensive operation before extracting back over the same bridge, having poked a stick into a hornets' nest, was going to be quite some undertaking. We rehearsed over and over again and I did have a contingency plan if the bridge got hit, but in fact we would probably have had to swim back.'

With Nimrod footage producing a positive intelligence picture, a pre-emptive strike could be ordered, as the Rules of Engagement required identification of enemy and a total assurance of the lack of civilians. That was now possible; each image collected in TAC HQ had shown armed Taliban fighters in 'Objective Vodka' with every building identified as having bunkers, defensive positions and firing slots. The plan for *Glacier 5* was to 'disrupt in the area of Vodka in order to protect the DC and deny any infiltration to cut us off'. Zulu Company would cross the river using the infantry assault bridge, stay on the far side of the canal for as long as it took to draw the enemy into a killing ground of the marines' choice, and then get out again. To do this in the dark among unrecced compounds, Andy Lock would rely heavily on his commanding officer for, on the ground, he had no means of seeing the Nimrod's down-feeds, which would be beamed into IX Group's command post's screens in the Garmsir DC.

The plan called for a diversionary attack by an Afghan National Police force just south of JTAC Hill, and an Afghan National Army artillery barrage into an area one kilometre to the north of the objective, in the hope that this would divert attention, if not men. Zulu Company set off at 2100 on 7 March with the Vikings stopping short of the canal, their progress constantly monitored by the HQ team on the video feeds from the circling Nimrod.

The first stage called for 59 Squadron's sappers to silently (and unobserved by the enemy) manoeuvre the bridge's many sections into position. At the best of times this would be a tricky and delicate operation, due to the twenty-metre width and need for silence; but in the Afghan spring, the canal was fast-flowing thanks to meltwater down from the mountains away to the north. The engineers struggled to lock each section of the bridge into place against the current, for as it reached closer to the far bank, the bridge – although stayed by ropes upstream – swayed alarmingly. Those watching stood in silence as they contemplated moving over this flimsy crossing, but once it was in place against the opposite bank and pinned down it locked into a firm unit and the marines waiting to cross were able to breathe again. The bridge, though, was always going to be the weakest link, so it was vital that it remained intact for the withdrawal – thus breaking a firm military principle of not returning via the same route. The Royal Marines on *Operation Glacier 5* had no alternative.

As the sappers were about to complete their job, the next phase of the plan unfolded while the marines watched a series of air strikes aimed at their primary objective. The aircraft made their bombing runs, dropping thousands of pounds of munitions onto the target – one 2,000-pound bomb hit an ammunition dump inside the compound, which offered an unexpected bonus, for the vast explosion not only forced a terrifying sound-and-light show into the heart of the enemy's camp but also threw tonnes of sand and rubble into the air. With so much dirt about, creating a fog, no one – least of all the Taliban – could see more than a few feet ahead. Using this cover, Zulu's marines ran across the bridge to take up their pre-planned defensive positions on the enemy bank. Andy Lock was delighted that the crossing went as well as it did – better than he'd feared – but as the troops approached their first objective they came under heavy and accurate fire; the tinderbox had been lit.

As soon as he was secure on the enemy bank, Lock and his teams began their move eastwards towards the known enemy positions with the major tucked in at the front, just one troop spread ahead of him. In a tight formation, so as not to lose track of each other in the chaos and darkness, the men cautiously approached the first compound. The fires burning all around from the bombing must have lit them up, though, because as they neared the walls the Taliban inside started firing on them. The marines immediately dropped into firing positions and shot back, the barrage so intense that the Taliban failed to see the troops separating and taking up different positions, as they started the crucial stage of the operation.

Silently, the men of 4 Troop, commanded by Second Lieutenant Matt Barlow with Sergeant Kenny Everett, moved into an over-watch position just short of 'Strongbow'. The hamlet was to be split in half with 1 Troop under command of Second Lieutenant Alec Ballard and Sergeant Taff Morris taking the north. 5 Troop with Second Lieutenant Matt Hammond and Sergeant 'Spinner' Spence would take the south. Each troop was to coordinate its movement with the other: they were advancing at similar speeds.

All the while, Rob Magowan was offering reassurance to his company commander: 'You are fine. You are fine. Don't worry. Can't see any enemy movement at all'.

Lock calmly reported the situation and was ordered to 'keep going, keep going', for the plan required all the known enemy to be lured into a central killing ground; then, on cue as it were, they now began 'pouring into their well-defended positions from the north, east and south'.

On agreement between the troop commanders, 1 Troop broke into the first compound, quickly followed by 5 Troop, with both lobbing grenades over walls and into every likely enemy hide as they went. Reaching the far side of this objective without much resistance, the men lowered their night-vision goggles into place and quickly scanned the open ground leading towards Vodka's more formidable defences. Nothing appeared to block their path, so 5 Troop readied themselves for the dash across while 1 Troop lifted their weapons to their shoulders and squinted down their sights, preparing to lay down suppressing fire should 5 Troop require it.

The company commander gave the order and 5 Troop dropped over the compound wall and headed across. As they did so, isolated pockets of Taliban soldiers started firing towards them; 1 Troop pinned the enemy down and called in artillery fire, at the same time dropping their own mortar bombs on the firing positions. As 5 Troop drew closer to their objective, the enemy, who had been holding much of their fire, took on the marines with a vengeance and a barrage of RPGs. But the deed had been done. The Taliban knew they had been outsmarted by the river crossing and were now contained in a neat target zone. On Robert Magowan's orders the F-18 aircraft circling overhead began their strafing runs.

It was time for Zulu Company to take stock. Not for the first time Lock noticed a strange phenomenon: 'The lads sometimes found this quite difficult to comprehend, for they go through their training doing exercise after exercise where they attack a building until those inside put the flag up and it's all over – victory! But it doesn't happen like that. It was going to be very costly to do that: for us to try and clear through that large area.

So much less costly to simply drop bombs on it, having forced the enemy to show his hand. That was the good news, but sometimes the lads felt "seen off" as they were always itching for a good personal scrap.'

Colonel Magowan, watching the individual movement of each Taliban, was confident enough to order Lock to start withdrawing before he was cut off. The inward leg was as successful as the outward journey and the marines of Zulu Company were back on their side of the river safely, watching the engineers collapse and retrieve the bridge before the whole raiding party returned to the District Centre by 0100 on 8 March. Thanks to superlative fieldcraft and intelligent fire-and-movement, there were no casualties.

With the enemy regrouped into Vodka's extensive compound – almost totally as far as could be ascertained by the Nimrod footage – the IX Group's ISTAR assets could concentrate on the Taliban gathered there. The desired 'target array' had been created by Zulu Company: it was up to the Apaches and B-1B bombers to finish the job. Most of the enemy had, for unknown reasons – but one that delighted Robert Magowan – run into the one building; twenty minutes later an F-18 Hornet bombed it, producing an explosion so large that 'it must have been another ammunition dump'.

With that success behind them, but not complacent, the colonel and his battery commander, Nick Sargent, spent the next four or five hours chasing a very few escaping enemy by video as they ran off towards the south, 'scattering all over the place'. ICOM chatter the next day suggested that twenty-nine Taliban lay dead in the former ammunition dump.

Those who had either not made it to the compound or who had, miraculously, escaped injury, were hounded by a Predator unmanned aerial vehicle. Four hours after the original attack, the 'pilot' announced that he had spotted another six Taliban ten kilometres south of Vodka. They were successfully engaged by the 105mm guns of 29 Commando Regiment, Royal Artillery, bringing *Operation Glacier 5* to a close.

For the next six days all was quiet in the Garmsir area.

On 23 March Colonel Magowan, whose Garmsir successes had impressed the Afghanis in Lashkar Gah, launched *Operation Malachite* in response to Helmand's civilian governor's view that as the Taliban had taken Babiji – to the north – they were about to threaten the regional capital itself.

The local ANA commander confirmed the seriousness of the threat by stating that, once lost, he was unable to take back the town with his available manpower. In a direct response to this apparent impasse, Magowan surged India Company to Lashkar Gah, secured the governor's compound, was given 42 Commando's Juliet Company, an Estonian Company and a company of the Worcestershire and Sherwood Foresters – who had just arrived in theatre – and set about restoring the status quo.

As it happened, the operation turned out to be a welcome anti-climax as Magowan, with this potent force, passed through Lashkar Gah before setting off on a two-hour drive north, where he found 'hundreds of ANA' and thought, 'What is going on here?' Backed by 'loads of Apaches, loads of men, Vikings, the lot', he pushed his teams towards Babiji through the dangerous *bocage* country of the green zone – 'no doubt full of Taliban' – with the ANA forward. 'We were right behind them as they literally just sauntered down the roads. The Taliban must have seen the force that was against them and cleared off.

'We stayed behind them the whole way as they cleared through the area. This time, though, it was only because Wafa, the provincial governor and the local ANA commander, were happy for us to do so. The Afghanis swept through the area in a short afternoon, whereas in Garmsir they weren't interested. Perhaps Garmsir was not their area for poppy. In the end that is what is going to have to happen. The Afghan solution.'

The next day Juliet Company was involved in a more serious action with a deliberate operation at Zumbelay, twelve kilometres to the east of Gereshk. As it was known to be a 'really bad place', Ewen Murchison was determined to seek a *shura* with the village elders. He knew the Taliban would be in evidence but as the civilians had asked him to visit, it seemed the ideal excuse for expanding his company's footprint, to take the initiative and to keep the enemy guessing.

To Juliet Company, Zumbelay held an almost mythical status. United States Special Forces described it as the gateway to the Sangin Valley; 3 Para had had a scrap there and an OMLT and ANA patrol had been caught out there. Juliet had tried to launch patrols but were, for one reason or another, always forced to turn back. Every Royal Marine knew that if they went in, it would lead to a contact. It was also one of the many places within the green zone where no intimate Viking support was available thanks to the numerous irrigation trenches that criss-crossed the area. On foot, the marines would carry considerable extra ammunition, knowing, too, that any wounded would have to be carried out.

Well short of the village two troops of Murchison's men – between forty-five and fifty marines – dismounted to begin their patrol forward when, almost instantly, women, children and old men began running in the opposite direction. Everyone recognised this clearest of all battle indicators. The expected contact was not far off.

Two sections of marines were in open ground when the Taliban started firing on them. The men dropped to the ground while a third section, under Sergeant Nige Quarman, pulled into some good cover, a crumbling brick wall at the edge of one of the fields above an irrigation ditch from where Quarman quickly directed his

GPMG gunner, Gav Teece, to provide covering fire by spraying the Taliban's firing positions with 7.62 ammunition. He then put Sam Vocea on mortar duty, and the Fijian cheerfully pumped 51mm bomb after bomb accurately on top of these same positions. Just behind was Second Lieutenant 'Dickie' Sharp accompanying an ITN film crew. Quarman's men's work at suppressing a number of the firing points was much to the film crew's relief.

The Taliban were using ten firing positions as well as engaging the company from the far side of the Helmand River to the west. For the press it was an unwelcome baptism, forcing Sergeant Quarman, during a lull, to suggest that he pull the ITN crew out altogether so that the marines could assault, unencumbered, into the compounds to their front. 'So, with covering fire, Bill Neilly and his overweight cameraman did an impressive 100-metre sprint into safety.'

Juliet Company's Afghan interpreter, Jamal, followed them, with rounds splashing around him until he fell, slumped against a bank. 'Bollocks,' Quarman muttered to himself, 'now I've got to run out and rescue him.' Luckily for the troop sergeant, Jamal had merely fainted. In tears and quickly recovered, he stumbled on, but not before Quarman remembered that 'he was just a civilian from Kabul, not a soldier, and these guys are really quite brave coming out with us.'

The exposed leading sections managed to dash into cover, regroup and continue the assault with Company TAC catching up as they all pushed forward to consolidate in the target compound – all under mortar fire. With the marines now firm on their objectives, the Taliban began withdrawing, but not before Marine Teece had killed many of them as they fled across a neighbouring compound. The enemy, though, was not yet finished, for indirect small-arms and RPG fire was being returned, with one rocket landing close to Marine Steve Atkinson. Covered in dirt, he stumbled towards the company's second-in-command, gesturing at the mortar line and yelling, 'Fucking hell, Boss, you gotta sort out that lot, just sort them out.' He then turned and stomped back to his position, still believing that the RPG that had so nearly killed him was 'one of our mortar bombs'.

With the defiant Taliban withdrawing deeper into the green zone and the village of Pasab, two kilometres to the north-east, 107mm rockets continued to fall among Juliet Company's position. 1 Troop assaulted the compound 2 Troop had suppressed by fire, to unearth from the battle debris a number of dead Taliban. The company now had no more ammunition with which to push further and, as it was officially a combat recce patrol or even a mere 'raid', it was time to return to the Vikings under a supporting artillery barrage. The Armoured Support Troop had, in the meantime, been engaged in their own stand-off against passing Taliban but the Vikings had retaliated with their GPMGs with, thankfully, no lasting damage. Murchison's marines re-embarked and returned to FOB Price. With clear indications that the enemy were reinforcing from the north, the decision to pull back had been wise.

'We were engaged,' Murchison reported, 'by quite a heavy weight of fire that heralded a two-hour battle. We cleared a few compounds, killed a number of Taliban and then pulled out. To me that was the sort of template I had learned over time. Spread ourselves out, prevent being outflanked, have good ICOM to understand what they are doing, keep mobile and always remain capable of getting out quickly in case of casualties.'

8
APRIL: SECURE –
DEFEAT – DESTROY

'The fools, they are firing into the wrong one, we are in the compound behind the one they are shooting.'

Long pause.

'Ahhh! They have now killed Sayid!'

Throughout much of the commando brigade's tour a 'cessation of hostilities' had existed in Sangin: this had been progress of a sort, even if ceasefires tended to benefit the Taliban more than the Royal Marines.

However, during the penultimate month of the deployment, Sangin's Taliban had again revealed their true intentions in waves of attacks against the District Centre, killing three men in a matter of days. As a parting blow, Brigadier Jerry Thomas decided that his brigade would assault the town and, once and for all, clear it of enemy. If that was successful, then one of his most enduring legacies for his successors would be a peaceful Sangin.

Perhaps there was something personal here too; the men didn't like to leave the DC, knowing they'd lost comrades there, without feeling like they'd achieved something as well.

The brigade battle group's 'taskorg' for *Operation Silver* included a battalion of the United States 82nd Airborne, 42 Commando's three Close Combat companies, plus Charlie Company RRF, now in

Sangin District Centre and about to be joined by Kilo Company, pre-positioned in advance of the operation, as well as available supporting units of gunners and sappers. Jerry Thomas's outline plan had the US battalion flying to the south of Sangin while the commando was, initially, to poise in the north-east, hoping to persuade the Taliban that that was the principal direction of attack. Meanwhile Matt Holmes had already planned a 'break-in'.

Holmes's Battle Orders were short and precise: Juliet, Kilo and Lima Company Groups were to 'Clear and Secure' Sangin and 'Defeat' any enemy they encountered 'within boundaries'; Charlie Company Group – while retaining over-watch from the District Centre – was, likewise, to 'Retain/Secure/Defeat' and the ever-present C Squadron, Light Dragoons, was to 'Screen/Secure' from the north-west through the north to the east. The Squadron would also escort the guns of 29 Commando Regiment, Royal Artillery.

On 4 April, 82nd Airborne landed by helicopter in the green zone, three kilometres south of Sangin, and began advancing north, clearing compounds on their way. Understanding the area, Holmes suspected they would take longer than planned and so, deploying, predominantly in Vikings, to the north-east, to deceive the enemy and disrupt his lines of communication, the colonel remained prepared to attack into Sangin. Foreseeing this possible chance, he and his industrious operations officer, Major Alex Janzen, had prepared the 'warning order' before the operation had even begun. Now their prescience was justified, for the 82nd had indeed become embroiled in a thorough, compound-clearing operation to the south, forcing the Taliban, who believed this to be the main assaulting force, to concentrate their fire accordingly. An enemy decision that gave Holmes the opportunity for which he was waiting. Now he could seize the initiative and strike directly into the heart of the town.

As the brigade commander, forward-based in the Sangin District Centre, was 'keen to get on with things' by increasing further the pressure on Sangin from the north-east, he directed Holmes accordingly. Holmes reflected on this latest order. Having suffered four mine strikes moving into position the day before, three of which resulted in nine casualties, many serious, and with the likelihood of taking more while manoeuvring on the outskirts of Sangin, all the while with two major manoeuvre units closing with each other from opposite directions – there had already been American infringements of 42's battle-space – Holmes called the commander back. Always one for exercising an offensive spirit, he suggested that he could 'break in tonight if given the resources – Nimrod MR2, Apaches etc. etc.'. The brigadier, keen to maintain the momentum, readily approved.

Jerry Thomas 'obviously thought the plan would work and the timing was right', so on the advice of his battery commander, Major Chris Lucas, Royal Artillery, Holmes further deceived the enemy by adopting the same fire plan he had used to suppress the Taliban's positions the night before when Kilo Company had flown into the District Centre. As the commando sped into Sangin, ICOM intercepts revealed that the enemy were indeed expecting more helicopters to resupply the DC. With their attention successfully diverted in the opposite direction, Holmes could now reinforce the deception effect by mirroring the previous blocking of Route 611 by Juliet Company; but this time, instead of the company sitting boldly on Route 611 to cut that line of communication, he moved Lima Company plus his own commando TAC HQ further south. Now all elements drove south 'straight down six kilometres of dirt track – Route 611 – right into the heart of Sangin in a rapid thrust that sought to surprise the enemy through shock effect. This track runs all the way from Kajaki to Gereshk and everyone who had ever been on 611 before had been ambushed, including French Special Forces and 16 Air Assault Brigade. It was quite a notorious route: very close country with high-walled compounds along its length, so it was not a route of choice. There were few, if any, run-offs; once we were moving we were fully committed. I knew damn well that I just had to move straight down it – and fast.'

Many of the marines – probably every marine – had his heart in his mouth, for, two weeks from the end of their tour and with home in their sights, they were now driving through an area where they'd been suffering mine strikes, down a road notorious for ambushes, straight into the general carnage of Sangin.

The convoy set off but, although everyone was expecting to get hit on the way in, either the subterfuge at the DC, where the Taliban were led to believe that the men were coming in on Chinooks, succeeded or the Taliban's leadership was focused on the American activities in the south. Either way the break-in was largely unopposed – compared with earlier deliberate actions – with the marines arriving in the centre of Sangin and forming up around the District Centre with no problems.

Meanwhile, down south, Holmes had been proved right: the 82nd was stuck, so it was time for 42 Commando to continue as they had started. When the commando reached the DC it about-turned to clear outwards with Lima Company heading south, Juliet Company north and Kilo Company pushing past Juliet to exploit further to the north-east.

Once more the marines were faced with the draining task of clearing compounds; only this time they had the benefit of numbers on their side, and the knowledge that, with the Americans moving up through the southern outskirts of Sangin, the Taliban were going to feel not only outnumbered but also squeezed. Instead of facing an enemy planning on outflanking them and cutting off their route back to base, this time the marines knew the boot was on the other foot; the Taliban would not want to find themselves wedged between two highly aggressive and tactically adept forces.

During this operation Corporal Thompson in his WMIK was, as always, lead vehicle for Juliet Company and was 'brilliant' at popping round the compounds, giving fire support when needed. While the Vikings had proved their worth in transporting the men to and from contacts, in a secure environment within its armoured walls, many of the men preferred to be in a WMIK as these vehicles offered a better all-round situational awareness, superb manoeuvrability and excellent firepower. Corporal Thompson in particular relished supporting the sections as they moved forward, using all the benefits of the WMIK in a pepper-potting and roving role. 'Somehow he was always there when you wanted him; giving instant, on call, fire support all over the place.'

The men moved forward in stages, making sure that they reached each of their objectives in turn, hopefully unscathed. When each 'line of exploitation' was crossed, defensive positions were established in buildings using large grain sacks to manufacture temporary sangers. Juliet Company established itself in the tallest building within its sector, placing the Fire Support Troop in an observation post on the roof. Corporal Hall's section, plus a 51mm mortar, became company reserve, to assist if anyone was threatened, while 3 Troop established defences around the governor's house, which was still under sporadic fire.

During the day the ICOM chatter, as so often, produced some almost surreal moments: 'The fools, they are firing into the wrong one, we are in the compound behind the one they are shooting.'

Long pause.

'Ahhh! They have now killed Sayid!'

By the time the commando had finished, the Taliban had been wiped from the area of the District Centre and 'out of the environs of Sangin'.

While direct contacts had been mercifully few – there were certainly enough to satisfy even the most 'offensively spirited' – over a three-hour period, Ewen Murchison, during the approach to Sangin, had four of his vehicles – one Viking, including his own, and three WMIKs – blown up by anti-tank mines. Fortunately, 'his' mine clipped the Viking's side with the force blasting upwards, removing all the bar amour. 'An inch further in and who knows what would have occurred; when it happens to your own vehicle it really concentrates the mind, especially for a company whose main task was to drive around in areas that had never been visited before.'

Although never 'good news', these mine strikes were a heartening endorsement for the new WMIKs with their underbody armour and V-shaped bellies. 'They stood up pretty well, with all of the guys walking away from them.'

Sangin, now at last all but empty of enemy, was described as being 'like a film set with the town utterly devastated'. To add to that devastation, a house-by-house clearing operation was conducted with 59 Independent Commando's engineers blowing up known firing points and sealing off and collapsing the underground rat-runs.

Operation Silver was a brilliantly executed and sudden finale for Holmes's 42 Commando. Relieved in place by 82nd Airborne 'who finally linked up with us', and, with no further casualties in what were his final days, he handed over responsibility and flew out of the country. 'A very quick end for us all, which was fantastic.'

Although a well-calculated gamble by Brigadier Jerry Thomas, *Operation Silver* was a success and a fitting conclusion to his brigade's tour. It was also a last reminder to the Taliban that the commandos had never been dissuaded from taking the war into the enemy's heartlands. Not for the first time the Royal Marines had treated the Taliban's presence with the contempt it deserved. *Operation Silver* opened up the Sangin Valley and set good conditions for the handover of 3 Commando Brigade to 12 Mechanised Brigade, who could then start expanding outwards to prosecute good deeds in the Sangin Valley.

With the departure of brigade headquarters and the major units, Andy Lock's Zulu Company became, briefly, a 12 Brigade unit for an operation in the Glacier series. *Operation Kendle* took place between 13 and 14 April with a new Fire Support Team and Mortar Fire Controller. Although now under 12 Mechanics Brigade, Lock noticed no great changes until he came under command of the new Light Dragoons squadron, 'then procedures noticeably altered'. 'During the operation we were aiming for a village but we didn't even get to the bloody place as we were caught up in a massive weight of fire and ended up in a three-hour daytime contact. It was difficult, particularly as we suddenly didn't have the old teams, such as the FST, who knew how to control the air and supporting weapons without guidance. A new team comes in to join the old and you expect them to drop ordnance and control Apaches and handle all those things that we had taken for granted up to then.'

9
MAKE SAFE

'... everyone fully embraced the commando quality of cheerfulness under adversity. This was best highlighted during three days of continuous rain, where Garmsir turned into Glastonbury (minus the music ... plenty of opium, though); when going on to check my vehicle I found the whole troop (minus the female signaller) naked, mud-wrestling ...'

The Transfer of Authority from 3 Commando Brigade to 12 Mechanised Brigade was completed on 11 April 2007 and the men were gradually on their way home. For some, this meant an opportunity, as they awaited transport out of Camp Bastion, for a long-awaited inter-section volleyball competition, after which they flew on to Cyprus for a few days of what one described as 'decompression'. For Corporal Alex Heath, returning to Bastion meant an end to 'the best weeks I have had in the corps yet; however,' he added, 'I now sleep soundly at night.'

As the brigade arrived back in the UK, His Royal Highness, Prince Philip, Captain General of the Royal Marines, insisted on greeting as many as he could at Exeter Airport: a much-appreciated 'welcome back'. His Royal Highness then visited Headquarters Commando Forces on 18 July – the day before the honours and awards were announced – and earlier, in a less publicised, moving and much-appreciated gesture, had supported Marine Benjamin Reddy's family during his funeral at the Royal Chapel, Windsor, on 19 March.

Reflecting on their months away was not an easy task, to explain all they had been through to those who hadn't been there a near-impossibility. 'Life on the front line?' said a marine. 'There is no front line. Life at the moment consists of a seven-month roller-coaster ride of emotions – a "normal" lifetime's experience – being ridden daily.' The analysis of their time away had to be done, though, in part to account for the men they'd lost, but also to prepare the ground for the next tour in Helmand Province. Officers and senior NCOs were in addition obliged to write citations for awards and generally mull over 'lessons learned'.

The most difficult of these tasks was, as ever, the honours and awards for bravery, with the Royal Marines continuing a reputation for understating their achievements. Corporal Cowe recognised that there were 'many individual acts of bravery', but pointed out that winning medals was not their aim, for 'you must be alive to wear them'. The bar for bravery during *Operation Herrick 5* was exceptionally high and there was a quota, and yet so many men did so many extraordinary things. CSM Shepherd felt that all his 'lads' had performed outstandingly: 'They have taken the intensity of the fighting fully in their stride and balanced it with sound judgement and restraint when needed.'

There is nothing so great as a ribbon but also nothing so potentially divisive. Much time and great care were expended by the brigade staff to ensure that the men of the UK Task Force of *Operation Herrick 5* were accurately recognised. Captain Graham Watson knew that it was the unity of purpose and spirit within his men that mattered: 'Throughout, no matter how intense the battle, mundane the job, overstretching [the] driver's hours (forty-eight hours non-stop on one occasion), or barking the plan, everyone fully embraced the commando quality of cheerfulness under adversity. This was best highlighted during three days of continuous rain, where Garmsir turned into Glastonbury (minus the music ... plenty of opium, though); when going on to check my vehicle I found the whole troop (minus the female signaller) naked, mud-wrestling ...'

Many believed that every man on the wings of the Apaches at Jugroom Fort should have been awarded a Military Cross; and while no Victoria Crosses were considered, the deeds of one or two marines were studied very closely.

When the awards were announced – the day before the Duke of Edinburgh visited Stonehouse Barracks – WO1 (RSM) Si Brooks of 42 Commando faced a number of marines, on the parade ground, rehearsing for the ceremony. 'Right,' he shouted, 'I want the following men to fall in over here. I don't know what you have done but the CO doesn't look very happy.' And with that he marched his small party towards the Officers' Mess where Matt Holmes, among other COs, was waiting. As the unsuspecting recipients approached they could see their commanding officers walking up the steps into the Mess. Marine Bishpam – about to learn of his Military Cross – when asked later why he thought his CO wanted to see him, believed that he was 'in the fucking rattle, sir' and kept muttering, 'I was *sure* she was over sixteen.'

Upstairs in the Officers' Mess, envelopes were laid out on a table alongside bottles of champagne. The brigade's chief of staff called for silence and announced, 'Gentlemen, it is my enormous honour to tell you that Her Majesty has bestowed on you honours and awards for your recent gallantry. I'm going to give you an envelope and I want you to read it and talk to your mates here and your COs and RSMs who are also here. Take a bit of time and we'll explain what it is all about and what is going to happen and then we'll drink some champagne.' Making that statement was, he said later, 'one of the most brilliant moments of my career'.

One marine from Whiskey Company, 45 Commando, whose CO couldn't get down from Scotland, was asked by a staff officer if he knew what it was all about. He said, 'Not really, sir,' so the officer reminded him, 'Do you remember the day Marine Holland died and you were pinned down and you fought your way out?' He seemed surprised that anyone knew. The officer continued, 'Yes, we do know all about that and it was pretty brave stuff, Marine Danby, which is why the Queen has given you a Military Cross. You should be really proud.' He was – and rightly so.

Brigadier J. H. Thomas had set out with very clear objectives at the outset of the tour, based firmly around supporting the Afghans in their efforts to rid themselves of the Taliban: 'The people will marginalise the enemy. I will magnify this by persuading and coercing the enemy into recognising the futility of his cause. Where necessary, I will defeat him; preferably I will render him irrelevant. I will manoeuvre to threaten, disrupt and interdict the enemy. He will be targeted on ground of my choosing at a time of my choosing. His defeat will be ensured but without enhancing his cause. These effects will endure through the development of credible Afghan security institutions for the long term.'

On their return, Brigadier (now Major General) Thomas was able to reflect on the success of the tour and especially on the professionalism and prowess of the men under his command: 'Whether it was 42 Commando in contact for thirteen hours during *Operation Slate* in November, the myriad patrol actions fought over several hours by our Mobile Operations Groups throughout Helmand, or the unrelenting attacks against our locations in Sangin, Now Zad and Garmsir, it was small groups and individuals who acquitted themselves with great courage, determination and professionalism in the highly dangerous situations they encountered repeatedly during our tour. Winter rains, flash floods and severe night-time cold combined with wide-open spaces interspersed with areas of densely packed compounds and thickly vegetated irrigation systems, plus an enemy inured to hardship, recklessly brave and motivated by extreme religious beliefs, require something special from our men and women who have to face them in combat. How superb they are and how proud I am to have served with them.'

Lieutenant Colonel Matt Holmes believed that 'the casualty rate between 16 Brigade and 3 Commando Brigade was noticeably different; high casualty rates are not necessarily indicative of a successful tour and there is no glory in it. Our aim during the tour was not to risk alienation and collateral damage by killing every Taliban we came across. The fighters are replaceable. Our aim was always to target their commanders and their crew-served weapons and the foreign fighters. We could identify, through communications, enemy activity, which helped to save lives by preventing us going into ambushes. We could also tell whom we were up against by the accuracy of the mortar fire: if the first rounds were close and then almost immediately adjusted to hit us, it was probably a Pakistani team controlling the mortar fire.

'I wanted to make the Taliban irrelevant to the population rather than kill every one of them. We used CIMIC projects to try and reinforce this effect, to demonstrate our good intent to the populace and convince them that we and their government were the only way ahead and that they should ignore the Taliban, who could offer just violence and intimidation.

'I wanted to do more in Kajaki but did not have the resources. Intelligence was limited. I think I had just two pieces of intelligence in the six months from which I could actually go and do something. At the end we had taken the fight to the enemy, had more contacts at our choosing and initiation, and had not been sitting waiting to be attacked in Platoon Houses. I – we all were – was very blessed with the exceptional quality of my marines: without a doubt, courageous and always willing to get back into the fray.'

Lieutenant Colonel Charlie Stickland, from his central position as the brigadier's chief of staff and now commanding officer of 42 Commando, sums up *Herrick 5*. 'The fighting had been intense, the risks considerable and the conditions demanding. In the beginning we unfixed ourselves across Helmand; we managed to establish redevelopment and target the insurgents in a more manoeuvrist fashion; we intensified the ADZ by establishing the DCCs and worked more closely with the Afghans. We took Babaji, the land between Lashkar Gar and Gareshk, and when we left it was free of Taliban, so you could drive through it or walk your dog through it. We deepened the ADZ rather than spread it out too quickly, otherwise it would have dried up. We set the conditions for the Kajaki dam project and, our last action under Jerry Thomas, we took Sangin on *Operation Silver*. Before that we owned the District Centre but not the town. To do that we delayed two companies going home and even called 59 Squadron back from the airhead as they were about to fly out. It was a very successful operation, for now people can patrol from the District Centre with more FOBs arranged around it.

'We prevented Garmsir from falling; we interdicted south and cut Taliban lines of communications; not permanently, but it relieved pressure, deepened the ADZ and made it geographically continuous. Musa Qal'eh continued to bubble and fester but we could never really have invested it. We believe we created manoeuvre, which is about all we could have done with one light brigade. Next time we go back, in autumn 2008, we will have three manoeuvre units plus an OMLT.'

Despite criticism voiced at home, the only thing the marines really missed were more marines. They could never understand why, when a commando's strength is above 700, it was capped by the government at 555. 'Do the politicians really want us to win this war and, if they do, then why don't they let us have our full quota of men? Perhaps even more than our quota to take account of casualties and those on leave? A section should be at least eight strong but ours were six or fewer and if one of those is on leave and then one gets wounded and it takes four to lift a stretcher, then that's the end of a section as an effective fighting unit.'

In the end, though, the offensive fighting spirit of the commando-trained, intelligent young men with their determination, superb motivation and extreme fitness – when coupled with the ability to manoeuvre – won the day.

The last word should perhaps go to the marine who, in reflecting on the months he and his fellow Royal Marines had spent in Afghanistan and the hardships they suffered together, felt a comradeship that couldn't be explained to anyone who wasn't there: 'Every man and his dog can look each other in the eye and understand what it means to wear the "green beret" and to have been part of *Operation Herrick 5*.'

Or there is a more succinct view: 'Will we miss it? Of course, so we will be back on *Herrick 9* in October '08!'

AFTERWORD

ADMIRAL SIR JAMES BURNELL-NUGENT, KCB, CBE
COMMANDER-IN-CHIEF FLEET

As the person in full command – and to make my own assessment of all of the operational, logistic and personal issues – I visited 3 Commando Brigade while on pre-deployment training. I saw them extensively in theatre. I visited families while on their own and I met some of the returning units. I also personally awarded a large number of commendations and awards: below Palace level.

By any standard this was an exceptional deployment in three key dimensions:

First, the very difficult ROE circumstances of not being at war but none the less in very hostile situations.

Second, the real-time close attention of the media – embedded or otherwise.

Third, when combat did occur – and pretty frequently it did too – it was intense, frightening, lethal, lasting and against an enemy who did not fight conventionally in terms of fear for his own life.

For every man to excel at every level of leadership and in combat is a resounding demonstration of the strength of the Royal Marines' capability in the conceptual, physical and moral components of a fighting power. At the time of their deployment, which coincided with the first operational deployment of 800 Naval Air Squadron at Kandahar with Harrier GR7s, as well as a large number of augmentees across all disciplines, 52 per cent of the United Kingdom troops deployed in Afghanistan were from the Naval Fleet Command. All of them, not just the elements of 3 Commando Brigade, performed to my glowing satisfaction as their commander-in-chief. My respect for the skill and capability demonstrated daily, and my personal congratulations for such success, are unbounded.

GLOSSARY

Of necessity this list is lengthy; fuller descriptions of many of these entries will be found in the text.

16 AAB	16 Air Assault Brigade
ADZ	Afghan Development Zone
AK-47	Kalashnikov 7.62mm (short) assault rifle with a range of 800 metres
ANA	Afghan National Army
ANAP	Afghan National Auxiliary Police
ANP	Afghan National Police
AO	Area of operations
Apache	Attack helicopter (AH-64) flown by the Army Air Corps
ASC	Armoured Support Company (Vikings), Royal Marines
ATV(P)	Viking all terrain vehicle (protected)
B-1B	USAF variable-wing, long-range 'Lancer' bomber capable of operating above 30,000 feet, delivering a wide range of guided and unguided munitions
Banyan	Picnic on a tropical beach – beneath a banyan tree
Barmine	Rectangular anti-tank mine used in Helmand to blast an entry through an obstruction
BDA	Battle damage assessment
Bde	Brigade: commanded by a brigadier
BG	Battle group: commando- or battalion-sized all-arms group under a single command
Bootneck	Slang for a Royal Marine
BPT	Brigade Patrol Troop, part of BRF
BRF	Brigade Reconnaissance Force

C2	Command and control
.5-cal	.5 inch, belt-fed, Browning M2 heavy machine gun carried by WMIK and Chinook. Effective range 1,850 metres
Cdo	Royal Marines commando of about 700 men (reduced to 555 for H5)
3 Cdo Bde	3 Commando Brigade, Royal Marines
CF	Coalition forces
CGC	Conspicuous Gallantry Cross
CH-47	Chinook helicopter
CIMIC	Civil-military cooperation
CO	Commanding officer of a commando/battalion-sized unit: lieutenant colonel
Coy	Company: up to four troops including a Fire Support Troop. Four in a commando, each up to 150 men commanded by a major
CP	Checkpoint or command post
CQMS	Company quartermaster sergeant
CSG	Command Support Group of HQ 3 Commando Brigade: was Headquarters and Signal Squadron
CSM	Company sergeant major
CTCRM	Commando Training Centre Royal Marines
CTR	Close target recognition
CVR(T)	Combat vehicle reconnaissance (tracked)
DC	District Centre or Compound
DCC	District Coordination Centre
Desert Hawk	UAV – model aeroplane – launched by a huge rubber catapult held by two men while a third pulls the aircraft back. The engine starts when it reaches a speed of 15 metres per second (50ft/sec). Supplies real-time video feeds back to a monitor. Endurance about one hour

DfID	Department for International Development
Dicking	Expression from Northern Ireland: to be watched and reported by non-aligned – but usually hostile – civilians
DIT	Development and influence team
DSO	Distinguished Service Order
Dushka	Soviet 12.7mm equivalent of the .5-cal.
ECM	Electronic counter measures
EOD	Explosive ordnance disposal
E-WMIK	Enhanced WMIK
F-15	USAF fighter, ground attack aircraft
F-18	USAF fighter, ground attack aircraft
FAC	Forward air controller
FARP	Forward air refuelling point
FCO	Foreign and Commonwealth Office
FFD	First field dressing
FLET	Forward line of enemy troops
FLOT	Forward line of own troops
FOB	Forward operating base
FOO	Forward observation officer, Royal Artillery
FSCC	Fire support coordination centre
FSG	Fire Support Group
FST	Fire Support Team
Galley	Royal Navy/Marines term for kitchen
GMG	Grenade machine gun firing 40mm grenades with a range up to 2 kilometres
GPMG	7.62mm, belt-fed general purpose machine gun. Range up to 1,800 metres
GPS	Global positioning system

H4	*Operation Herrick 4* (16 Air Assault Brigade)
H5	*Operation Herrick 5* (3 Commando Brigade, Royal Marines)
H6	*Operation Herrick 6* (12 Mechanised Brigade)
H9	*Operation Herrick 9* (3 Commando Brigade, Royal Marines)
Harrier GR7	RAF and Royal Navy (during H5) vertical take-off, ground attack aircraft
HE	High explosive
Heads	Royal Navy/Marines term for latrines
HEG	Helmand Executive Group
Hellfire	Air-to-ground anti-armour, fire-and-forget missile system used by AH against fortifications and compounds. Range between 500 metres and 8 kilometres
Hexamine	Small solid-fuel tablets for heating water and food
Hilux	Toyota, hybrid, four-wheel-drive saloon with a 'pick-up', open back; often containing Taliban in the front and a mortar base-plate and other heavy weapons in the back
HLS	Helicopter landing site
HMG	Heavy machine gun
Honking	Royal Marines slang, antonym to hoofing
Hoofing	Royal Marines slang for 'excellent'
HRF	Helmand Reserve Force
HSSG	Helmand Security Steering Group
HUMINT	Human intelligence
ICOM	Hand-held VHF used by the Taliban
IED	Improvised explosive device
ILAW	L21A1: 84mm, one-shot, unguided anti-armour weapon. Range 300 metres
IRT	Immediate response team

ISAF	International security assistance force
ISF	Indigenous security forces
ISTAR	Information, surveillance, target acquisition and reconnaissance
IX	Information exploitation
JARIC	Joint Air Intelligence Reconnaissance Centre
Javelin	Fire-and-forget, anti-tank missile system used in the 'bunker-busting' mode
JDAM	Joint direct attack munition
JDCC	Joint District Coordination Centre
Jingly truck	Brightly-coloured, heavily ornamented, Afghan civilian, heavy-lift lorries
JNCO	Junior non-commissioned officer: lance corporal and corporal
JOC	Joint Operations Centre: providing the infrastructure for 42 Commando's battle group headquarters
JPCC	Joint Provincial Coordination Centre
JTAC	Joint terminal attack controller
Kandak	Afghan battalion
KIA	Killed in Action
King's Badge	King George V on 7 March 1918 directed that the senior squad of Royal Marines recruits under training should be titled 'The King's Squad'. He subsequently approved that the most outstanding recruit should be awarded 'The King's Badge'. This badge, the cypher GvR within a laurel wreath, is worn throughout his service by a King's Badgeman, no matter what rank he may later reach and in all orders of dress

.338 LRR	Sniper rifle. Accurate range 1,100 metres
LD	Light Dragoons
LD	Line of departure: start line for a deliberate attack
LMG	7.62mm magazine-fed light machine gun. Range 1,100 metres. (Taliban)
LMG	Minimi, light machine gun (UK)
LUP	Lying-up position
MC	Military Cross
MCF	Military construction force
MFC	Mortar fire controller
MiD	Mention in despatches
Minimi	5.56mm, belt-fed light machine gun. Range 1,000 metres. (UK)
Mne	Marine
MOB	Main operating base
MoD	Ministry of Defence
MOG	Mobile Operations Group
51 Mor	51mm light mortar: HE range of 750 metres
81mm Mor	81mm mortar: HE range of 5,650 metres. Also fires smoke and illumination
120mm Mor	120mm heavy mortar used by the Taliban. Range of 7,240 metres
Mouse hole	X-shaped charge used to blast a hole through an obstruction
MR2	Nimrod: RAF reconnaissance and surveillance aircraft
NATO	North Atlantic Treaty Organization
NGO	Non-governmental organization
Nimrod	RAF MR2 reconnaissance and surveillance aircraft
OC	Officer commanding units smaller than a commando/battalion

ODA	Operational Detachment Alpha (US)
OEF	*Operation Enduring Freedom*
OGD	Other government departments
OMLT	Operational mentoring and liaison team
OP	Observation post
Ops One Coy	Reserve rifle company at instant notice as a QRF

3 Para	3rd Battalion, the Parachute Regiment
Pathfinder	16 AAB equivalent of the BRF
Pinzgauer	Small, light, troop carrier with a tubular framework body
PJHQ	Permanent joint headquarters
PKM	7.62mm GPMG. Range of 1,500 metres
PRR	Personal role radio carried/worn by every marine. Range of 500 metres
PRT	Provincial Reconstruction Team
PSCC	Provincial Security Coordination Centre
PSYOPS	Psychological operations
PVCP	Permanent vehicle checkpoint

QIP	Quick impact project
QCVS	Queen's Commendation for Valuable Service
QRF	Quick reaction force

R&D	Research and development
R-and-R	Rest and recuperation leave (fourteen days after six months in theatre but including travelling time)
RC(S)	Regional Command South
RCIED	Radio controlled improvised explosive device
RIP	Relief in place
ROC	Rehearsal of concept (ROC drill)
107 Rocket	107mm Chinese anti-armour rockets. Range in excess of 7,000 metres

ROE	Rules of engagement
RPG	Rocket-propelled grenade: Russian-made anti-tank missile more often used by terrorists. 90mm calibre with a maximum (aimed) range of 500 metres and capable of penetrating 260mm of armour.
RSM	Regimental sergeant major; senior non-commissioned officer within a commando
RSOI	Reception, staging and onwards integration
RV	Rendezvous
SA-80	5.56mm standard infantry rifle. Optimum range 300 metres, effective to 600 metres
SAF	Small-arms fire
Sangar	Small, defensive position surrounded by a stone or sandbagged parapet
Sappers	Nickname for the Royal Engineers
SAS	Special Air Service: UK army special forces
SATCOM	Satellite communications
SBS	Special Boat Service: predominantly Royal Marines, special forces
Sec	Section: three or four to a Royal Marine troop; between six and eight men strong, usually commanded by a corporal
SF	Security forces or special forces
SH	Support helicopter
Shura	Meeting – usually with the elders of a town or district
SIED	Suicide improvised explosive device
SIGINT	Signals intelligence
Sitrep	Situation report
SNCO	Senior non-commissioned officer: sergeant, colour sergeant, warrant officer second class and warrant officer first class

SOP	Standard operating procedure
Sqn	Squadron
Stripey	Royal Marines sergeant: usually a troop sergeant
TAC HQ	Tactical headquarters of a unit, e.g. commando TAC HQ
TACP	Tactical air control party
TF	Task force
TI	Thermal imaging
TIC	Troops in contact
Toms	Soldiers in the Parachute Regiment
Tp	Troop (Royal Marines equivalent to an army platoon): about thirty strong commanded by a second lieutenant, lieutenant or junior captain
Troop Bible	Generic term for a small notebook kept by the troop sergeant or CSM, containing brief details of every individual: including service number, blood group, rifle number, name of next of kin and so on
UAV	Unmanned aerial vehicle
UKTF	UK task force
UOR	Urgent operational requirement
VBIED	Vehicle-borne improvised explosive device
VCP	Vehicle checkpoint
Viking	ATV(P)
WIA	Wounded in action
WMIK	Weapons mount installation kit (fitted to a stripped-down Land Rover)

APPENDIX ONE

THE ROYAL MARINES

Since their formation, the Royal Marines have been part of the Royal Navy and, with the corps's deployment in Afghanistan during the 2006/2007 winter, 52 per cent of all British forces in theatre – over 400 miles from the nearest sea – were from the naval service.

On 28 October, 1664, 1,200 land soldiers of the Admiral's Regiment were divided into six companies each of 200 men and distributed into His Majesty's fleets and prepared for sea service. Over the following 343 years the number of Royal Marines has fluctuated: during the Second World War it was as high as 78,500, while the current figure is just over 6,500, or 20 per cent of the Royal Navy. Apart from the units of the 3rd Commando Brigade, detachments and individuals of every rank serve in a number of Her Majesty's Ships and submarines as well as around the globe in a variety of roles: some on exchange or loan service and others as observers and liaison officers with international organisations in, at the last count, 31 different countries.

At 32 weeks, Royal Marines recruit training is longer than officer training in some countries and is conducted at the Commando Training Centre, Lympstone. The Commando Course – universally acknowledged as the hardest infantry course in the world and which, for recruits, is the penultimate stage of their basic training – is aimed at the mind as much as the body of both officers and marines and is spent entirely in the field: uniquely nobody returns to camp throughout. It is a progressive course that, in addition to the physical demands forced upon the candidates, places great emphasis on mental alertness when at the extremes of exhaustion.

It culminates in a final eight-day exercise designed to test officers and marines in all aspects of the training they have so far undertaken and which is particularly testing and realistic, incorporating a raid from the sea, an abseil and an assault. On arrival back at Lympstone the candidates face the Commando Tests themselves: an Endurance Course, Tarzan and Assault Course, a Nine Mile Speed March and a 30 Mile Load Carry, which take place one a day without a break. On successfully completing the commando tests, the officers and marines are awarded their Green Berets, mindful of the four elements of the 'commando spirit' that will have been imbued in them: courage, determination, unselfishness and cheerfulness in the face of adversity.

While other individual military units may claim not to have seen much active service since the Falklands war of 1982, units of Royal Marines have, over the intervening 25 years, been employed on operations in Northern Ireland, the Lebanon, Hong Kong, Belize, both Gulf Wars – elements remain deployed in Iraq – south-east Turkey, Honduras, Nicaragua, the Balkans, Sierra Leone, Cyprus, Afghanistan and the Congo; and as if that was not enough the commando brigade continues to deploy to north Norway each winter for Arctic training. It is hardly surprising, therefore, that His Royal Highness, Prince Philip, the corps's captain general, is on record as saying: 'Nothing is impossible for the Royal Marines.'

APPENDIX TWO

HONOURS AND AWARDS

DSO

Lieutenant Colonel Matthew John HOLMES Royal Marines
Brigadier Jeremy Hywel THOMAS Royal Marines

MBE

Major Oliver Andrew LEE Royal Marines
Acting Lieutenant Colonel Ewen MURCHISON Royal Marines
Warrant Officer Class 2 Martin Charles REEVES Royal Marines

CGC

Corporal John Thomas THOMPSON Royal Marines

MC

Marine Matthew BISPHAM
Marine Daniel CLARICOATES
Corporal Michael COWE Royal Marines
Marine Ian Paul DANBY
Marine Daniel FISHER
Corporal Alan HEWETT Royal Marines
Sergeant Jason LAYTON Royal Marines
Corporal Simon WILLEY Royal Marines
Staff Sergeant Keith John ARMATAGE Army Air Corps
Lance Bombardier Richard David JENNINGS Royal Artillery
Captain David Charles RIGG Royal Engineers
Warrant Officer Class 1 Mark Simon RUTHERFORD Army Air
    Corps

MiD

Captain Alistair Scott CARNS Royal Marines

Marine Thomas CURRY (killed in action)

Captain Duncan Graham FORBES Royal Marines

Major Philip GADIE Royal Marines

Marine Jonathan Paul HART

Marine Alexander HOOLE

Warrant Officer Class 2 Robert Daniel JONES Royal Marines

Corporal Adam LISON Royal Marines

Marine Paul MAYFIELD

Captain Jason Robert MILNE Royal Marines

Corporal Ashley OATES Royal Marines

Second Lieutenant Richard SHARP Royal Marines

Warrant Officer Class 2 Steven SHEPHERD MBE Royal Marines

Marine Matthew SMITH

Sapper Lee Stephen HORAN Royal Engineers

Corporal Steven MARNER Royal Engineers

Warrant Officer Class 2 Michael SMITH Royal Artillery (killed in action)

Major Jonathan SWIFT Royal Regiment Fusiliers

Sergeant Michael WILKINSON Light Dragoons

QCVS

Major Matthew CHURCHWARD Royal Marines

Lieutenant Colonel Duncan Andrew DEWAR Royal Marines

Lieutenant Colonel Charles Richard STICKLAND Royal Marines

Major Benedict John WARRACK Light Dragoons

AUTHOR'S NOTES AND ACKNOWLEDGEMENTS

This is not an analytical account of 3 Commando Brigade's deployment in Helmand Province on *Operation Herrick 5*, nor is this a history book. First, it does not and cannot cover the whole of the six-month tour of duty, for there were so many individual skirmishes that there is simply not the space. Secondly, due to the arcane restrictions of the Official Secrets Act, another thirty years need to pass before the formal release of facts and figures can produce a balanced account.

This book, therefore, offers a discrete snapshot of events seen through the eyes of some of those who took part in at least 54 dedicated or enduring operations, let alone the dozens upon dozens of minor – but no less dangerous – firefights and 'scraps'. For instance, 42 Commando alone had over 520 contacts – not counting the numerous isolated mortar bombs or RPG rocket attacks 'to which we became quite oblivious' – while 821 deliberate engagements were recorded across the brigade: nearly twice as many as their predecessors. These eyewitness accounts are presented through vignettes culled from a few individuals who served with 3 Commando Brigade during that Afghan winter of 2006–2007. Nor is it a straightforward, chronological narration – although I have tried – for much was happening, concurrently, across every corner of the brigade's area of operations.

I am, though, deeply conscious that I have paid less tribute to some units than others – space and time being my lame excuse. In particular I wish I had been able to devote considerable passages to the skill and professionalism of 29 Commando Regiment, Royal Artillery; 28 Engineer Regiment, Royal Engineers; 131 and 59 Independent Commando Squadrons, Royal Engineers; and the equally noble – and equally indispensable – members of the vastly diverse Commando Logistic Regiment.

I specifically acknowledge the 'unsung hero' status of the men, mostly from 45 Commando, who formed the Operational and Mentoring Liaison Teams across the province: their work was at the very heart of the brigade's mission and yet it so often took place far from the mis-named 'glamour' of the battlefield. Working with, and alongside, members of the Afghan National Army, they faced considerable danger and professional frustration, yet never once wavered from their aim. The success of *Operation Herrick 5* was very much their success: they more than need a book of their own. Coupled with this, there has been little space to record the less kinetic achievements of the commando brigade as a whole: the numerous schools, hospitals and maternity clinics that were constructed and opened across the area of operations.

Nor have I been able to record the remarkable deeds of the medical staff whether in the back of a Chinook flying-in a T1 casualty or in the field hospital at Camp Bastion. I hope the deeds of all these people will be presented to a wider public: they deserve it. The pilots in valiant support – on call and willing at every hour of the day and night that flying was possible, in their B-IB bombers, Nimrods, F-18s, F-15s, Apaches, Lynx or Harriers – need their story to be told too.

At home, one day, someone will tell, publicly, of 'Yorkie' Malone and his enthusiastic supporters – many from Tavistock in Devon – and their remarkable 'shoebox campaign' that dispatched, and continues to dispatch, parcels of essential 'goodies' to the sailors, marines, soldiers and airmen – and women – on Afghanistan's front line.

In putting this saga together – in just over five months since signing the publisher's contract – I have been helped by so many. Some I will mention here, while others will be aware of my appreciation in the text. However, if anyone has been inadvertently left out, I apologise and will address the error in a subsequent edition.

Before I present my general list of acknowledgements, I must thank His Royal Highness, Prince Phillip, Captain General of the Royal Marines, for honouring 3 Commando Brigade's Helmand account with his generous foreword.

I am also grateful to General Sir Richard Dannatt, Chief of the General Staff, for allowing me to quote him on the title page.

In the beginning Brigadier Mark Noble, Director, Royal Marines, answered my initial request with a positive and encouraging response. These days no 'civilian' – indeed no serviceman – will be given 'official' permission to publish an account of a continuing operation, while serving personnel can only be interviewed 'with permission' – and even then they may not, for private reasons, wish to take part.

My next stop was the brigade commander himself, now Major General Jerry Thomas, who, likewise, encouraged me 'unofficially' to pursue the project: from then onwards, and with help from two announcements in the Royal Marines journal the *Globe and Laurel*, a floodgate opened.

Many officers, SNCOs, JNCOs and marines – spurred on no doubt by their commanding officers – were willing to assist. Lieutenant Colonel Matt Holmes, lately of 42 Commando, set the scene so eloquently over a two-hour session when he certainly had more important affairs on his mind. But I had always to remember that 42 Commando was not the only manoeuvre unit in the brigade. The brigade's chief of staff, Lieutenant Colonel Charlie Stickland, and his operations officer, Lieutenant Colonel Oliver Lee, gave me much time from their ridiculously busy schedules as did Lieutenant Colonel Duncan Dewar, commanding officer of 45 Commando, and Lieutenant Colonel Robert Magowan of the Command Support Group, aka the Information Exploration Battle Group. Major Ben Warrack of the Light Dragoons was unstinting in his support, while Admiral Sir James Burnell-Nugent was kind enough to offer a high-level explanation of one or two contentious aspects of the campaign.

Others who helped with a similar willingness include, in no particular order: Major Martin Collin, Captain Tony Forshaw, Lieutenant Colonel Andy Maynard, Corporal Mick Cowe, Lieutenant Paul Youngman, Royal Navy, Captain David Rigg, Royal Engineers, Corporal Alan Hewett, Major Steve Fraser, Marine Mark Farr, Sergeant Darren Stubbings, WO1 (RSM) Tony Jacka, Sergeant Jay Layton, Corporal Neil Chappell, Major Andy Lock, Major Jim Grey, Captain Chris Witts, Captain Jason Milne, Major Ewen Murchison, Major Neil Sutherland, Major Steve Liddle, Corporal Mick Slunker, WO2 (CSM) Steven Shepherd (now RSM of 45 Commando), Sergeant Nige Quarman, WO2 (CSM) Marty Pelling (now RSM of 40 Commando), Corporal John Thompson, WO1 (RSM) Si Brooks, Acting Sergeant Russ Coles, Marine 'Bungy' Williams, Lance Corporal Jim Wright, Corporal Jamie Sanderson, Acting Corporal Mike Gregson, Lance Corporal Gav Teece, Sergeant Fergie Gask, Sergeant 'Spinner' Spence, Sergeant 'Kenny' Everett, Corporal Al Weldon, Marine Adam Edwards, Corporal Paul Fillingham, Major Robby Money, Irish Guards, Lance Corporal Kieran Scott, Marine Phil McDougall, Corporal Tyler Alldis, Corporal Chris Gardiner, Lance Corporal Jim Spencer, Father Michael Sharkey, Royal Navy, Corporal Bugler Dan Johnson, Captain Rajiv Joshi, Royal Army Medical Corps, Captain Mike Davis-Marks of MOD PR (RN) and Colonel Ben Bathurst of MOD PR (army).

Ian Ballantyne of *Warship International Fleet Review* helped with photographs, while Captain John Hillier, editor of the *Globe and Laurel*, allowed me to quote, plagiarise and copy!

I asked Tim Mitchell – a Royal Marine, signaller for ten years – of TM Design to draw the maps and so help him establish his new business by working on something familiar. I am delighted with the result and wish him well.

I would like to thank the Presidents of the Sergeants' Messes at the Commando Training Centre, Lympstone, and 42 Commando at Bickleigh, plus the President of the Mess Committee of the

Officers' Mess at Stonehouse for their hospitality and the use of rooms for interviews.

As always, photographs pose a problem and on this occasion I have received over 5,000 stills and a great many video clips; many of which were sent 'anonymously'. Clearly some of these photographs in private collections were originally from official sources but sent to me with no accreditation. At least three official Royal Naval photographers were with the brigade and to whom credit must certainly be given: Petty Officer Photographer Sean Clee, Royal Navy, Leading Airman Photographer Andy Hibberd, Royal Navy, Leading Airman Photographer Gareth Faulkner, Royal Navy, while Marine 'Buster' Brown of 7 Troop, Lima Company 42 Commando took a few. The photographs of the Apaches at Jugroom Fort were taken by Sergeant Gary Stanton, Royal Air Force.

Robin Wade, my agent with an interesting military background himself, has, as he has had to do so often, kept me to schedule and 'to contract'. His wisdom when we needed to alter this account from a military history into something more appealing to a wider audience, was invaluable. In this regard I thank Humphrey Price for his help in transforming the manuscript from 'niche market' to 'mass market'.

My family, led by Patricia, have suffered from my unusually tight deadline and borne the work schedule with considerable tolerance.

Finally, I thank again the senior officers of the Royal Marines who so willingly granted me unhindered access to their men and establishments, their personal thoughts and correspondence. Having said that, opinions not directly attributed to any one person are mine, for which I take full responsibility, while the views of those interviewed do not represent those of the Ministry of Defence.

A proportion of the royalties from *3 Commando Brigade* are being donated to Royal Marines charities and the *Help for Heroes* campaign.

Ewen Southby-Tailyour
South Devon, Spring 2008

BOOKS BY EWEN SOUTHBY-TAILYOUR

Military history/biography
Falkland Islands Shores
Reasons in Writing: A Commando's View of the Falklands War
Amphibious Assault Falklands: The Battle of San Carlos Water
Blondie: A Life of Lieutenant-Colonel HG Hasler, DSO, OBE
The Next Moon: The Remarkable True Story of a British Agent
    Behind the Lines in Wartime France
HMS Fearless: The Mighty Lion

Fiction
Skeletons for Sadness

Reference
Jane's Amphibious and Special Forces
Jane's Special Forces Equipment Recognition Guide

INDEX